Monographs in Virology

Vol. 8

Editor: J.L. MELNICK, Houston, Tex.

S. Karger · Basel · München · Paris · London · New York · Sydney

Virus Infections in Bats

S. Edward Sulkin and Rae Allen

Department of Mibrobiology,
The University of Texas Southwestern Medical School at Dallas,
Dallas, Tex.

1 figure and 14 tables

19 74

S. Karger · Basel · München · Paris · London · New York · Sydney

Monographs in Virology

Vol. 4
NEURATH, A. ROBERT and RUBIN, BENJAMIN A. (Philadelphia, Pa.):
Viral Structural Components as Immunogens of Prophylactic Value
VI + 87 p., 11 fig., 1 cpl., 1971
ISBN 3-8055-1203-1

Vol. 5
WILDY, P. (Birmingham): Classification and Nomenclature of Viruses
VIII + 81 p., 1971
ISBN 3-8055-1196-5

Vol. 6
TINSLEY, T. W. and HARRAP, K.A. (Oxford):
Moving Frontiers in Invertebrate Virology
VIII + 66 p., 23 fig., 3 tab., 1972
ISBN 3-8055-1464-6

Vol. 7
BECKER, Y. (Jerusalem): The Agent of Trachoma
XIV + 99 p., 12 fig., 5 tab., 1974
ISBN 3-8055-1657-6

S. Karger · Basel · München · Paris · London · New York · Sydney
Arnold-Böcklin-Strasse 25, CH-4011 Basel (Switzerland)

Contents

Editor's Preface

Work on this Monograph was nearing completion when the sudden death of Dr. SULKIN intervened. However, having worked closely with Dr. SULKIN for nearly 20 years and having been involved from the beginning in accumulating the data for this book, Ms. ALLEN graciously agreed to complete the work. It is a fitting commemoration to Dr. SULKIN to have the contributions he made in the area of viruses and bats drawn together into a single volume.

In this Monograph are presented data concerning natural and experimental virus infections of bats, as well as ways in which experimental data can be projected into the planning of field studies. Perhaps the single most interesting aspect of viral infections in bats is the apparent lack of host response to established infection. This phenomenon may lead to large numbers of healthy carriers of rabies virus in bat populations, which maintain ever-shifting foci of infection and thus provide a constant source of infection for man and animals. Bats also act as an ideal reservoir host for arboviruses, providing infective blood meals for vectors. One indeed must be vigilant for the role that bats may play in the maintenance and dissemination of viruses to other animals and to man.

JOSEPH L. MELNICK

Acknowledgments

A number of individuals played significant roles in the planning and execution of the studies originating from this laboratory. Without their contributions this work would not have been possible. The authors are indebted to Dr. PHILIP H. KRUTZSCH, mammalogist-anatomist, who participated in the initial phases of these studies and advised concerning the capturing and handling of bats and to Dr. RUTH A. SIMS, Dr. STANLEY K. TAYLOR, Dr. WILLIAM E. STEWART II, Dr. BETTY A. HATTEN, Dr. LARRY L. LEONARD, Dr. BOBBIE L. MIDDLEBROOKS, Dr. E. PABLO CORREA-GIRON, Dr. CHANSOO KIM and Dr. YONG-JIN YANG for participation in various aspects of the program.

To those who assisted in collecting bats in widely scattered areas of the United States, Mexico and Japan the authors are extremely grateful. These include Prof. HAROLD B. HITCHCOCK, Bates College, Lewiston, Maine; Prof. JAMES B. COPE, Earlham College, Richmond, Indiana; Prof. BERNARD VILLA-R., Institute of Biology, National University of Mexico; WOODROW GOODPASTER, taxidermist, Cincinnati, Ohio; WILLIAM HANSZEN, rancher, Burnet, Texas; F. GLEN ANDERS, University of Houston; TEX VILLARREAL, Corpus Christi-Nueces County Health Department, Texas; Dr. SHOICHI IIDA, Dr. SEIICHI TOSHIOKA and others at the 406th Medical Laboratory, US Army, Japan; Dr. HIROO IIDA, Dr. KEISAKU HATTORI and Dr. NORIO SAKURADA of the Hokkaido Institute of Public Health; Dr. KIYOTOSHI KANEOKO and Dr. KANJI MIYAMOTO of the Tokyo Medical and Dental University, and Dr. TAKEO FUKUDA and Dr. TOHRU SASAHARA of the Miyasaki Institute of Public Health, Japan. Dr. JORDI CASALS of the Yale Arbovirus Research Unit, New Haven, Connecticut, was most helpful in the initial phases of the identification of certain virus isolates. Special thanks are due RAY CHRISTIAN, LA RUE CHRISTIAN, HIROKO KUDO, YUMIKO KAMIJO and TERU UDAGAWA for diligence in the maintenance of bats in the laboratory and for technical assistance.

These investigations were supported by the Lampert Foundation of Beverly Hills, California; research grant 5 R01 AI02316 and training grant

5 T01 AI00142 from the National Institute of Allergy and Infectious Diseases, United States Public Health Service; and National Science Foundation Grant GB-12611. Certain phases were conducted under the sponsorship of the Commission on Viral Infections, Armed Forces Epidemiological Board, and were supported by the US Army Medical Research and Development Command, Department of the Army under contract No. DA-49-193-MD-2138, and under grant No. DAHC 19-68-G-0008 from the Life Sciences Division, US Army Research Office.

Introduction

In many instances, the fact that certain diseases of man may afflict other species as well was realized long before the infectious agent responsible for the disease was identified. In ancient writings hydrophobia in man was associated with the bite of mad dogs; the belief that rodents were carriers of plague was exemplified by a rat-like animal in the foreground of a 15th-century etching of St. Quirlin, the plague saint [HULL, 1955]. The search for reservoir hosts of human disease-producing agents among domestic and wild animal populations has now progressed until it forms a major objective in modern epidemiological methods. Over the years man has learned that, in order to combat human disease, steps must be taken to recognize the causative agents in animal populations and to attempt to control these natural reservoirs. Of the many groups of animals, both wild and domestic, which have been involved in epidemics, the mammalian order Chiroptera is perhaps the most recent group to gain widespread attention as a reservoir of infectious agents transmissible to man.

The first documented evidence of a virus disease transmitted by bats occurred in 1908 in Santa Catarina in southern Brazil, although the paralytic disease in cattle, horses and mules (called *mal de caderas*) was not diagnosed as rabies until 1911 [CARINI, 1911] and the bat was not proved to be the vector until 1921 [HAUPT and REHAAG, 1921]. For many years interest in bats as carriers of rabies virus was confined to South and Central America and Mexico where the frugivorous and vampire bat species implicated in the transmission of this agent are native. The demonstration of natural rabies virus infection in insectivorous species of bats native to the United States [IRONS et al., 1954; SCATTERDAY, 1954; SULLIVAN et al., 1954; VENTERS et al., 1954; WITTE, 1954] provided the impetus for numerous field and experimental studies designed to evaluate the potential role of bats as reservoir hosts for rabies virus and, subsequently, for certain arboviruses. A compilation in 1962 of documented evidence suggesting that bats could serve as effective reservoir hosts for these viruses was persuasive [SULKIN, 1962] and during the past decade much additional information has accumulated associating bats with these and other viruses. The purpose of this monograph will be to present the known facts concerning natural and experimental virus infections in bats, to indicate the ways in which experimental data can and have been projected into the design and interpretation of field studies, and to offer conclusions as to the significance of these findings. Finally, since the virus-infected bat may show no overt signs of disease, special experiments designed to define the mechanisms responsible for the balanced virus-host cell interactions operative in these inapparent infections will be discussed.

Certain unique specializations of bats relating to their anatomical, physiological, behavioral and ecological characteristics are particularly meaningful with regard to the effectiveness of these animals as reservoir hosts for viruses. Recent additions to the literature concerned with various aspects of the biology of bats indicate increasing interest and productivity in this research area. BARBOUR and DAVIS [1969] have compiled excellent accounts of the life histories of bat species found in the United States. A collection of papers presented at a recent symposium devoted entirely to bats provides current information from the laboratories of those researchers most deeply involved in studies of bats [SLAUGHTER and WALTON, 1970] as do the first two volumes of a multivolume treatise exploring major facets of chiropteran biology [WIMSATT, 1970a, b]. In our brief discussions of certain characteristics of bats which relate to their effectiveness as reservoir hosts for viruses (i.e., flight, migration, habitats, food preferences, longevity, thermoregulation, and hibernation and the role of brown adipose tissue) we beg the indulgence of specialists in these areas for any errors in interpretation or possible superficial treatment of their very fine studies and hope always to have included pertinent references to guide those readers interested in learning more about these extraordinary mammals.

Characteristics of Bats Relating to their Effectiveness as Reservoir Hosts for Viruses

For a terse statement of the singularity of bats among mammals we find the words of JEPSEN [1970] particularly descriptive: 'Apparently no other grade of mammal has ever had so many eccentric and extremely "specialized" characteristics combined into such a highly "successful" organism.' Bats are among the most ancient of mammals. The earliest form yet discovered is from the early Eocene epoch. This specimen, found buried on the bottom of Fossil Lake in southwest Wyoming, possesses the morphological character- istics of a true bat and thus provides no clues to the ancestor of the bat. It is generally agreed that bats originated at some considerably earlier time and probably were derived from a line of arboreal insectivores. For informa- tion concerning bat origins and evolution the reader is referred to JEPSEN [1970], who described and named the earliest known fossil bat, *Icaronycteris index* [JEPSEN, 1966]. The history of the classification of bats as reviewed by PETERSON [1964] emphasizes the dilemma of the early taxonomists intent on finding a niche for these strange animals. First placed in a category designated simply as being different from the common rule, bats were subsequently classified with the primates by LINNAEUS who placed them in a single genus, *Vespertilio,* comprising 7 species. This obvious misplacement was subsequent- ly righted and the order Chiroptera (hand-wing) was established to encom- pass the numerous diverse forms of these unusual mammals. Bats were first arranged systematically on the basis of structure of the bony skeleton and on the development, structure and arrangement of the teeth. One of the most extensive early classifications was that of MILLER [1907]. In recent years, the development and application of new biological techniques for studying relationships among animals (i.e., karyotype analyses, serological tests, etc.) have tended to condense the classification of bats, and KOOPMAN and JONES [1970] suggested that the trend to reduce the number of recognized bat taxa at the generic and specific levels will continue. An abbreviated classi- fication of the order Chiroptera, modified after KOOPMAN and JONES [1970], is presented in table I. The order Chiroptera comprises 2 suborders, 16 cur- rently recognized families, about 170 recent genera and some 847 recent

Table I. Classification of bats [1]

Classification	Common names	Genera	Species
Suborder Megachiroptera			
Family			
Pteropodidae (3) [2]	fruit-eating, flying foxes	38	ca. 150
Suborder Microchiroptera			
Superfamily Emballonuroidea			
Family			
Emballonuridae (2)	sheath-tailed, sac-winged	12	44
Noctilionidae	fish-eating, bull-dog	1	2
Rhinopomatidae	mouse-tailed	1	2
Superfamily Phyllostomatoidea			
Family			
Phyllostomatidae (6)	leaf-nosed, vampire, fruit bats	48	ca. 123
Chilonycteridae	mustached, leaf-chinned	3	8
Superfamily Rhinolophoidea			
Family			
Megadermatidae	false vampire, yellow-winged	4	5
Nycteridae	hollow-faced, hispid	1	13
Rhinolophidae (2)	spear-nosed, horseshoe	11	ca. 128
Superfamily Vespertilionoidea			
Family			
Furipteridae	smoky bats	2	2
Molossidae	free-tailed	11	ca. 82
Mystacinidae	short-tailed	1	1
Myzopodidae	sucker-footed	1	1
Natalidae	long-legged, funnel-eared	1	4
Thyropteridae	disc-winged	1	2
Vespertilionidae (6)	simple-nosed	34	ca. 280

1 Modified after KOOPMAN and JONES [1970].
2 Number in parentheses indicates number of subfamilies.

species. Bats are second only to rodents in numbers of living genera and species and are the largest order of mammals in overall abundance. The Megachiroptera, or Old World fruit bats, are all considered to be a single family, the *Pteropodidae,* whereas the Microchiroptera, which exhibit a great deal more diversity, are divided into 15 distinct families.

Bats are more widely distributed throughout the world than any other mammal except man. They are believed to be primarily tropical animals, however, since, discounting the *Mystacinidae,* a small family confined to

New Zealand, all the families are found in the tropics whereas only 3 families (*Rhinolophidae, Vespertilionidae, Molossidae*) inhabit temperate zones as well [KOOPMAN, 1970]. Species within these families have adapted to temperate zones by developing the ability to hibernate during the winter months or the perception to migrate southward for this period of the year.

Of the several characteristics of bats which seem most significant in an evaluation of the importance of these animals as reservoir hosts for viruses, the fact that they are volitant is of prime consideration. Studies concerned with annual migrations and with movements within resident areas in nightly forages for food indicate that these animals would be capable of seasonal reintroduction of viruses into areas free of the agents for periods of time as well as dissemination of infectious agents within a given area [BARBOUR and DAVIS, 1969; DAVIS *et al.*, 1962; GRIFFIN, 1970]. Improved marking and banding techniques have provided the means for demonstrating migrations of several hundred miles [cf. GRIFFIN, 1970] and for establishing the fact that at least some of the small species weighing 4–6 g may have a life-span of over 20 years [EISENTRAUT, 1960; GRIFFIN and HITCHCOCK, 1965].

It has been theorized that bats evolved as night flyers to avoid competition with birds for insects. Research into the mechanism which enabled bats to fly in darkness began in 1793 with the studies of LAZAR SPALLANZANI who demonstrated the ability of blinded bats to avoid obstacles in flight and, acting on the suggestion by a fellow scientist, LOUIS JURINE, found that the sense of hearing was vital for the successful flight of these animals [cf. DIJKGRAAF, 1960]. Unfortunately, much of the work of these early scientists remained unpublished and more than a century elapsed before the true mechanism by which bats navigate in the darkness was realized. HARTRIDGE [1920] suggested that bats made use of sounds of high frequencies and short wavelengths to perceive objects at a distance; 20 years later, after the development of radar, the work of GRIFFIN and GALAMBOS completely defined 'echolocation' [GRIFFIN, 1958].

The food preferences of the various families and genera of bats are quite diverse. Probably the most familiar groupings according to food sources are the insectivorous species, the fruit-eaters, and the vampires, the only animals known to subsist entirely on a diet of blood. In addition, there are also carnivorous species which feed on small mammals, fish, amphibians and reptiles, as well as species which utilize nector and pollen as foodstuffs. For discussions concerning the evolution of various skeletal structures and dentitions as related to feeding habits of bats see GLASS [1970], SLAUGHTER [1970] and VAUGHAN [1970]. Comments concerning the food preferences and feeding habits of bats are germaine to a discussion of bats as reservoir hosts of viruses, particularly in regard to rabies virus and the arthropod-borne viruses as discussed below.

Although bats are most often thought of as inhabiting caves and caverns in remote places, many species are known to occupy man-made structures as daytime roosting sites [DALQUEST and WALTON, 1970]. Mexican free-tailed bats in particular are often found in close proximity to man. DAVIS et al. [1962] found this species roosting in buildings in 87 of 150 towns visited in south Texas and observed that age, type of architecture, relative position in a town or use by humans failed to exclude a building from possible habitation by these animals.

While most mammalian species have thermoregulatory mechanisms that prevent marked variations in body temperature, thermoregulation in Chiroptera is as diverse as the group itself [HENSHAW, 1970; LYMAN, 1970; DAVIS, 1970]. The Megachiroptera regulate body temperature as true homeotherms whereas microchiropteran species, depending on whether they inhabit tropical, subtropical or temperate zones, may exhibit patterns of thermoregulation which vary from constant homeothermy to less precise homeothermy to what appears to be complete abandonment of control in which body temperature drops to ambient whenever the animal ceases to move. The most recent studies on thermoregulation in bats indicate that with increasing knowledge in this area it becomes more dangerous to generalize on the subject and, particularly with temperate zone species, one must consider only data accumulated for an individual species in its natural habitat. For our purpose of relating body temperatures of bats to their role as reservoir hosts for viruses, however, we feel justified in stating that certain species which inhabit temperate zones may experience, without stress, variations in body temperature over a range which could not be tolerated by any other mammalian species. Bats in flight or those resting in an attic during the daylight hours in summertime may have body temperatures well above 37 °C. Cave bats may be subject to periods of daily hibernation when, after night-time feeding flights, they come to rest in the cool environment of their habitat and body temperatures drop to ambient. The most dramatic response of bats to temperature is observed in the deep hibernating species that experience prolonged periods of dormancy during the wintertime, when body temperatures may reach lows of just a few degrees above the freezing temperature of tissues. Although body temperatures of hibernating bats may rise periodically during dormancy, there is evidence that bats maintain low body temperatures and decreased metabolic rates for longer continuous periods than other hibernating mammals. Studies on the physiological bases of hibernation in bats and the mechanisms which control entrance into and arousal from hibernation form a large part of chiropteran literature [DAVIS, 1970; HENSHAW, 1970]. Bats appear to be unique among mammalian hibernators in their quickness to respond to low temperature and enter a torpid state and, also, in the speed with which they can become active again. For hundreds of years investigators probed the mysteries of the brown adipose tissue found most abundantly in hibernating animals and gave the organ-like accumulation of this fat in the interscapular region the name 'hibernating gland' because they believed it had a function relating to hibernation. JOHANSSON [1959] reviewed the numerous reports concerned with the physiologic function of brown adipose tissue and concluded that this tissue appeared to be involved in the regulation of body temperature; soon afterward firm evidence that brown fat is, indeed, a thermogenic tissue was published [SMITH, 1961]. Since that time several studies concerned with the role of this tissue as a site of thermogenesis and a thermogenic effector of arousal in hibernating animals have appeared [SMALLEY and DRYER, 1967; HAYWARD and LYMAN, 1967; SMITH and HORWITZ, 1969; LINDBERG, 1970] and it would seem that the early designation of interscapular brown fat as the hibernating gland is now justified.

Laboratory Care and Maintenance of Bats

Early investigators had found it difficult to adapt certain species of bats to life in captivity or found it possible to do so only under a regimen which would be impractical in studies involving infectious agents and large numbers of bats [KUMM, 1932; GATES, 1936, 1938; RAMAGE, 1947]. Prior to initiating studies on experimental viral infection in bats in our laboratory, methods had to be devised to maintain these animals in captivity in a healthy state for long periods of time. Formulation of a balanced diet which the insectivorous bat would accept in captivity was critical, as was the design of housing which would suit these animals and protect laboratory workers caring for large numbers of infected bats. In our initial studies, Mexican free-tailed bats *(Tadarida brasiliensis)* were used and a number of feeding techniques together with numerous food combinations were devised to induce these animals to take food without being hand-fed. A diet which proved highly successful in maintaining this bat species, and subsequently little brown bats *(Myotis lucifugus)* and big brown bats *(Eptesicus fuscus),* in a healthy state for long periods of time consisted of a homogenized mixture of cottage cheese, bananas, cod liver oil and mealworms [KRUTZSCH and SULKIN, 1958]. The roughage provided by the chitinous exoskeletons of the mealworms prevented packing and resultant intestinal obstruction. This was in keeping with a suggestion by GATES [1938] that chitin seems to be necessary to maintain a healthy alimentary condition. Subsequently, cod liver oil was deleted and liver extract with iron and a multivitamin preparation were added to the formula and dry cottage cheese substituted for regular large curd cottage cheese in accordance with a report by MOHOS [1961]. As a further modification, agar-agar in nutrient broth was added to the mixture in a final concentration of 1% to reduce the consistency to a semi-solid state [SULKIN *et al.,* 1963]. Bats which survive 7 or 8 days and learn to eat the food can now be kept for several months or even years. Of the 3 species used in our studies, the big brown bats adapt best to life in captivity.

For housing bats in the laboratory, a cage was designed which would allow the animals to position themselves as they do in various roosting places in nature and which would facilitate feeding and transfer of bats in and out of the cages with safety. The initial cages were rather elaborate, fitted with narrow access slots closed by sliding metal strips and a sliding floor panel for removal of dead animals [KRUTZSCH and SULKIN, 1958]. When we became more familiar with the way these animals behaved in captivity, a simpler model was designed. The cage currently in use consists of a wooden frame covered on the top, sides and back with stainless steel hardware cloth. The floor is solid wood as is the front, which is hinged at the bottom to open down and has a sliding glass panel in the center for observation purposes and through which bats can be introduced into the cage singly after virus inoculation. Bats become quiescent within a few hours after being placed in these cages, either upon first arrival at the laboratory or after subsequent virus inoculation, and have a tendency to cluster in the top rear area. Once animals have settled down in their new environment the front of the cages can be opened to introduce food and water in small disposable containers. Bats are maintained at different environmental temperatures for various experiments but always in darkness and at a humidity of about 60%; they are fed in the late afternoon to coincide with their normal feeding time in nature. The animals remain quiescent throughout the day but become active and move around the cages and vocalize as feeding time approaches.

Other investigators involved in studies of the role of bats in various infectious diseases or of various aspects of the biology of bats have devised diets and housing which have proved satisfactory for their purposes [CONSTANTINE, 1952; STONES, 1965; TESH and ARATA, 1967; SIMPSON and O'SULLIVAN, 1968; BARBOUR and DAVIS, 1969]. Techniques for the netting of bats are summarized by BARBOUR and DAVIS [1969].

Rabies Virus

Natural Rabies Virus Infection in Bats

The association of bats with rabies is probably the oldest and certainly the most thoroughly documented of the known relationships of these animals with virus diseases transmissible to man. This association was first recognized in southern Brazil in the early 1900s when a paralytic disease of cattle and horses was diagnosed as rabies, yet could not be correlated with canine rabies [CARINI, 1911]. Certain patterns in the epizootiology of the disease in herbivores suggested a flying sylvan host [HAUPT and REHAAG, 1921] and bats were investigated as transmitters based on reports from inhabitants of the affected districts concerning the abnormal behavior of bats, flying about and biting cattle during daylight hours. Interestingly, the first bats incriminated were fruit-eating *Phyllostoma,* whereas proof that vampire bats *(Desmodus)* were involved in the dissemination of rabies virus was not obtained until several years later [QUEIROZ LIMA, 1934]. Following the first reports from investigators in Brazil, the disease, or more probably its recognition as bat-transmitted rabies, seemed to move northward as similar information was obtained in other South American countries and in Trinidad, Central America and Mexico [ENRIGHT, 1956]. The first human deaths due to rabies transmitted by bats occurred in Trinidad in 1929 [PAWAN, 1936a]. Rabies infection in bat populations on the island reached epizootic proportions around this time and resulted in 89 human deaths, the last in 1937. The first case of vampire bat rabies in man reported outside Trinidad occurred in Mexico in 1951 [MALAGA-ALBA, 1954].

The prevalence of rabies virus infection in bat populations throughout the Latin American countries and the resulting livestock epizootics and human deaths had little impact in the United States because the ranges of the bat species involved did not extend into this country. There were, however, sufficient indications of the presence of vampire bats in association with rabies in southern California to warrant an investigation in 1952; positive proof that vampire bats were involved in the incidents occurring in this area was not obtained [ENRIGHT, 1956]. No attempts had been made to isolate rabies virus from any of the bat species native to this country, although one group of investigators, prompted by the observations in Latin America, had reported in 1951 that two species of insectivorous bats were susceptible to experimental infection with rabies virus [REAGAN and BRUECK-NER, 1951]. Also, in 1951, a human rabies death had occurred in Texas in which the case history contained a reference to the patient's having been bitten by a bat [SULKIN and GREVE, 1954]. Despite these suggestions that bats in the United States might be carriers of rabies virus, no investigations of this possibility were begun until after a young boy in Florida was attacked and bitten in June, 1953, by a bat subsequently shown to be infected with rabies virus [VENTERS et al., 1954]. A second human exposure to a proven rabid bat occurred in Pennsylvania a few months later [WITTE, 1954]. These episodes prompted public health officials in other states to examine insectivorous bats in their areas, and rabies virus was isolated from bats collected in Texas in November and December, 1953, and in California in July, 1954 [IRONS et al., 1954; SCATTERDAY, 1954; SULLIVAN et al., 1954; VENTERS et al., 1954]. The next state reporting bat rabies was Montana, where the virus was isolated from the brain of a bat submitted for examination in August, 1954 [BELL et al., 1955]. Thus, within about 1 year of the first isolation of rabies virus from an insectivorous bat in Florida, evidence was obtained of rabies virus infection in bats captured in 4 additional, widely separated states. Table II presents information concerning the emergence of bats as natural wildlife hosts for rabies virus in the United States, beginning with the first cases of bat rabies detected in 1953. The states are listed chronologically according to the year in which bat rabies was first reported; the cumulative totals of numbers of states reporting bat rabies appear parenthetically in this column. The total number of states reporting bat rabies each year is also shown along with a figure in brackets denoting the number of states in which bats were the only animals in which rabies virus infection was demonstrated during that year. The number of cases of bat rabies confirmed and entered into the records of the National Center for Disease Control each year is shown and related to the total number of cases of wildlife rabies for that year as the percent of these cases which bats constituted. Subsequent to the first isolation of rabies virus from bats in Florida, Pennsylvania, Texas, California and Montana in 1953 and 1954, additional states reported cases of bat rabies each year until, by 1967, evidence of bat rabies had been reported from all of the contiguous states except Rhode Island; in June, 1967, a rabid bat was found in this state, leaving only Alaska and Hawaii

Table II. Bat rabies: distribution and number of cases in the United States, 1953–1972.
Proportion of total wildlife cases composed of bats[1]

Year	States reporting bat rabies for first time	Number of states reporting bat rabies	Number of cases of bat rabies	Number of cases of wildlife rabies	Percent bat cases/ wildlife cases
1953	Florida, Pennsylvania (2)[2]	2	8	1,479	0.5
1954	California, Montana, Texas (5)	4 [1][3]	4	1,697	0.2
1955	Louisiana, New Mexico (7)	6 [1]	14	1,915	0.7
1956	Alabama, Georgia, Michigan, Minnesota, New York, Ohio, Oklahoma, Utah (15)	11 [2]	41	2,079	2.0
1957	Arizona, Colorado, Wisconsin (18)	11 [0]	31	1,942	1.6
1958	Nebraska (19)	11 [1]	68	2,075	3.3
1959	Connecticut, Illinois, Maryland, Virginia, West Virginia (24)	18 [2]	80	1,915	4.2
1960	Indiana, Iowa, Kansas, Missouri, New Jersey, Oregon (30)	17 [3]	88	1,836	4.8
1961	Arkansas, Kentucky, Massachusetts, South Dakota, Tennessee, Washington (36)	28 [3]	186	2,174	8.6
1962	Idaho (37)	27 [5]	157	2,314	6.8
1963	Delaware, North Carolina, Nevada, New Hampshire, South Carolina (42)	36 [8]	303	2,611	11.6
1964	Maine, Mississippi, Vermont (45)	36 [7]	352	3,560	9.9
1965	North Dakota, Wyoming (47)	42 [6]	484	3,257	14.9
1966		40 [7]	377	2,946	12.8
1967	Rhode Island (48)	43 [8]	414	3,211	12.9
1968		44 [9]	291	2,702	10.8
1969		38 [8]	321	2,672	12.0

Table II (continuation)

Year	States reporting bat rabies for first time	Number of states reporting bat rabies	Number of cases of bat rabies	Number of cases of wildlife rabies	Percent bat cases/ wildlife cases
1970		45 [11]	296	2,554	11.6
1971		47 [12]	465	3,449	13.5
1972		45 [17]	459	3,348	13.7

1 Data provided by Rabies Control Unit, Center for Disease Control, US Department of Health, Education and Welfare.
2 Figure in parentheses is cumulative total of number of states reporting bat rabies.
3 Figure in brackets refers to number of states in which bats were the only rabid animals reported.

not having reported the isolation of rabies virus from bats. The bat population of Alaska is sparse and no systematic surveys to detect rabies infection in these animals have been conducted; rabies infection has been demonstrated only in the wild and domestic canids of Alaska [RAUSCH, 1958]. In Hawaii, rabies has never been diagnosed in any animal species or in man [SULKIN, 1965]. The number of states reporting cases of bat rabies each year increased gradually after 1953 and the totals listed show the wide distribution of bat rabies throughout the United States, particularly from 1963 onward. Of special interest are the figures in brackets which indicate the number of states each year in which bats were the only rabid animals reported. During 1968 through 1971, for example, in 20–25% of the states reporting cases of bat rabies, bats were the only rabies-infected animals reported. This statistic was even more significant in 1972, when 17 of the 45 states reporting bat rabies would have listed no recorded cases of animal rabies were it not for the presence of rabid bats.

The number of cases of bat rabies detected in the United States each year increased steadily after the initial discovery of rabid bats in 1953. The question often posed is whether the figures on record represent true increases in the number of bats infected with rabies virus each year or if they are simply a reflection of the growing awareness of health officials and the general public of the existence of bat rabies. CONSTANTINE [1967], in an analysis of data concerning bat rabies in California since the time of its detection there in 1954, found that more rabid bats were detected each year because larger numbers were submitted for rabies tests in successive years; the percent of samples positive for rabies infection did not increase. Of 266 bats tested in California prior to 1960, 33 (12%) were infected with rabies virus, compared with 49 of 465 (10.5%) in 1964, and 72 of 653 (11%) in 1965. CONSTANTINE interprets these data as indicating that bat rabies may have been present in the United States long before it was detected in Florida in 1953.

Since the number of cases of bat rabies reported each year prior to around 1963 probably does not present a true picture of the extent of bat rabies in the United States during that period, the relationships of these totals to the total number of cases of wildlife rabies during that time are not as meaningful as those based on data accumulated in more recent years. Yearly totals since 1963 show that the number of states reporting bat rabies, the number of cases reported and the percentage of total wildlife rabies cases which the bat cases represent have remained fairly constant.

Bat rabies was detected in Canada in the summer of 1957 when the virus was isolated from a big brown bat *(E. fuscus)* taken from the Vancouver area in British Columbia [AVERY and TAILYOUR, 1960]. Since that time, natural rabies infection has been demonstrated in additional bat species in British Columbia and in Ontario and Manitoba [BEAUREGARD, 1969]. In the most recent survey covering the years 1963–1967 it was noted that the rabies virus infection rate was highest in migratory bats which winter in the United States, presumably in contact with bats of the same species known to be carriers of rabies virus. Also, most isolations in British Columbia and Ontario were from bats taken from the southern counties of the provinces, again suggesting contact with rabid bats from the United States [BEAUREGARD, 1969]. Review of the National Communicable Disease Center's Zoonoses Surveillance Reports on rabies for 1972, which included monthly summaries of rabies cases in Canada, indicates that the incidence of bat rabies in that country is small compared to the number of cases in other wildlife hosts.

Information concerning natural rabies infection in bat species in other areas of the world is much less extensive than that which has accumulated for the western hemisphere. The existence of rabies infection in bats in India was indicated by a report in 1954 concerning a man who died of rabies some 3 months after being bitten on the forearm by an unidentified bat [VEER-ARAGHAVAN, 1954]. Additional information concerning the extent of bat rabies in India has not come to our attention. Other countries in which rabid insectivorous bats have been found include Yugoslavia and possibly Hungary [NIKOLIC and JELESIC, 1956], Turkey [TUNCMAN, 1958], and Germany [PITZSCHKE, 1965; WERSCHING and SCHNEIDER, 1969]. The only report of rabies virus infection in a megachiropteran bat is from Thailand, where 2 of

79 dog-faced fruit bats *(Cynopterus brachyotis)* were shown to be naturally infected [SMITH *et al.*, 1967]. KAPLAN [1969] reports that fairly extensive surveys in Indonesia and in western and southern Africa have produced no evidence that bats in these areas are naturally infected with rabies virus. In a survey carried out in the United Arab Republic more than 400 bats, consisting of 9 insectivorous and 1 fruit-eating species, were collected from ancient Egyptian temples, pyramids, mosques, churches, caves and hollow trees; rabies virus was not isolated from pooled brain specimens nor was neutralizing antibody against this agent demonstrated in pooled serum samples from these animals [EL-SABBAN *et al.*, 1967]. Several of these bat species, however, were shown to be susceptible to experimental rabies virus infection.

Experimental Rabies Virus Infection in Bats

The first experimental studies on rabies infection in bats were those carried out by investigators in southern Brazil and Trinidad, intent upon proving that vampire bats were responsible for the epizootics of rabies in livestock in these countries [QUIEROZ LIMA, 1934; TORRES and QUIEROZ LIMA, 1935; PAWAN, 1936 b]. In those studies it was observed that while some experimentally infected bats developed furious or paralytic forms of rabies and were capable of transmitting virus, others showed no overt signs of illness yet were equally capable of transmitting the disease to other animals [PAWAN, 1936 b]. Thus, the experimentally infected vampire bat provided the first model of the symptomless carrier of rabies virus, a condition which was subsequently shown to exist in nature [TORRES and QUIEROZ LIMA, 1936].

Investigators in the United States, prompted by the reports concerning rabies and bats in Latin American countries, demonstrated that two species of insectivorous bats native to this country *(M. lucifugus* and *E. fuscus)* were susceptible to experimental infection with street virus strains of canine origin [REAGAN and BRUECKNER, 1951]. Subsequent to the first reports of the isolation of rabies virus from naturally infected insectivorous bat species of the United States [IRONS *et al.*, 1954; SCATTERDAY, 1954; SULLIVAN *et al.*, 1954; VENTERS *et al.*, 1954; WITTE, 1954] and the report of a human death from rabies believed to be due to bat bite [SULKIN and GREVE, 1954], several groups of investigators undertook studies concerned with determining the susceptibility of different insectivorous bat species to various strains of rabies virus and with attempting to demonstrate bite transmission of rabies virus from bats to other animals [REAGAN *et al.*, 1954; ENRIGHT *et al.*, 1955;

STAMM *et al.,* 1956; BURNS *et al.,* 1958]. A number of bat species were shown
to be susceptible to strains of rabies virus administered subcutaneously,
intracerebrally, intranasally or intramuscularly. In all these studies, infected
bats developed signs of central nervous system disease and died, or died in
the absence of previously observed symptoms. Rabies virus was demonstrated
in brains and salivary glands (the only tissues tested) and in the saliva samples by mouse inoculation and bite transmission to hamsters [REAGAN *et al.,*
1957] but not by bite transmission to monkeys or guinea pigs [BURNS *et al.,*
1958]. Incubation periods were often quite prolonged and symptoms, when
observed, consisted of tremors, signs of irritability and aggressiveness, and
paralysis. Most of these studies were carried out with small numbers of bats,
which were tested for evidence of infection only when exhibiting signs of
illness or when found dead. Thus, although rabies virus had been isolated
from the brains and salivary glands of apparently healthy, normal-behaving
bats taken from nature [SCATTERDAY, 1954; BURNS *et al.,* 1956], the symptomless carrier state was not demonstrated in the first studies with experimentally infected insectivorous bats.

In studies initiated in the authors' laboratory in 1956, experiments were
designed to characterize rabies virus infection in bats more definitively with
the view to learning more about the susceptibility of different bat species
to different strains of rabies virus and making observations on the course of
the infection, particularly with regard to the demonstration of virus in bats
which showed no overt signs of rabies. In considering various physiological
characteristics of bats which might be involved in the ability of these animals
to sustain persistent inapparent infection with rabies virus, attention was
drawn to the presence of interscapular brown adipose tissue in these animals.
Although at that time the presently known role of brown fat as a thermogenic
effector of arousal in hibernating animals such as bats and ground squirrels
[SMITH and HORWITZ, 1969] had not been defined, there was evidence that
this tissue served an important function relating to seasonal periods of dormancy. The high level of metabolic activity attributed to brown fat suggested
that this tissue would provide an ideal site apart from the central nervous
system for the replication and persistence of rabies virus in the inapparently
infected bat. The growth of coxsackievirus and poliovirus in the brown fat
of various laboratory animals had been reported previously [DALLDORF,
1950; PAPPENHEIMER *et al.,* 1950; ARONSON and SHWARTZMAN, 1956]. Thus,
studies were designed to characterize experimental rabies virus infection in
bats, with special emphasis on determining whether virus administered intramuscularly would invade and multiply in the interscapular brown fat of these

animals. In the initial studies Mexican free-tailed bats *(T. brasiliensis)* were used because a number of the first bat rabies virus isolations were made from this species and because tremendous populations of *Tadarida* are present in the southwestern part of the United States. Subsequently, little brown bats *(M. lucifugus)* were also obtained for study because brown adipose tissue is particularly abundant in this deep hibernating species. In addition to determining the susceptibility of bats maintained at room temperature to experimental infection with rabies virus, we took into consideration the unique thermoregulatory mechanisms of bats and studied the influence of environmental temperature on rabies infection in these animals. Also, because of the well-known effect of pregnancy on the susceptibility of various animals to virus infections, experiments were included to determine the influence of the gravid state on the course of infection in spring-breeding Mexican free-tailed bats.

Initial experiments concerned the susceptibility of Mexican free-tailed bats and little brown bats maintained at room temperature to a canine strain of rabies virus [SULKIN *et al.,* 1959]. Virus was administered intramuscularly and tissues obtained for assay included brain, salivary glands and interscapular brown adipose tissue harvested over a period of from 2 to 12 weeks following inoculation from animals which had been dead less than 4 h or were sacrificed either when showing signs of disease or in apparent good health. Evidence of infection was demonstrated in approximately 20% of the *Tadarida* and 40% of the *Myotis* included in the experiments. This strain of rabies virus exhibited neurotropic characteristics, and was demonstrated most frequently in the brain tissue of infected bats but was also found in salivary glands and brown adipose tissue. The levels of virus demonstrated in brown fat indicated that replication occurred in this tissue. Signs of central nervous system disease and death were observed in some bats of both species, and a large portion of the animals shown to be infected were those which were found dead or were sacrificed when signs of disease were evident. Bite transmission of rabies virus from a *Myotis* showing signs of irritability and aggressiveness to a suckling mouse was demonstrated during the course of these experiments [SULKIN, 1962]. Rabies virus was also recovered from the tissues of bats which showed no overt signs of illness at the time of sacrifice. Although *Myotis* appeared to be more susceptible to this strain of rabies virus than *Tadarida,* the overall results obtained with both species were similar. In a series of experiments with *Tadarida* inoculated intramuscularly with the canine strain and maintained at 24 °C, tissues from 492 bats were tested for the presence of rabies virus and evidence of infection was obtained

Table III. Demonstration of rabies virus (canine strain) in tissues of experimentally infected *Tadarida brasiliensis* maintained at 24 °C[1]

Condition of bats	Time after inoculation, days	Number of bats			
		tested	with virus demonstrated in		
			brain	salivary glands	brown fat
Dead <4 h or showing symptoms of rabies virus infection	18–66	69	62 (0.90)[2]	32 (0.46)	13 (0.19)
Apparently healthy	13–83	35	29 (0.83)	5 (0.14)	10 (0.29)
Totals	13–83	104	91 (0.88)	37 (0.36)	23 (0.22)

1 From Sulkin and Allen [1970].
2 Number in parentheses represents frequency of virus demonstration.

for 104 (21 %) of these animals. Table III shows the distribution of virus in the various tissues of 69 of the infected bats which were found dead or were sacrificed with symptoms 18–66 days after inoculation as compared to tissues obtained from 35 of the infected animals which were in apparent good health when sacrificed 13–83 days postinoculation. Virus was demonstrated in brain tissue of both groups with approximately equal frequency, whereas animals exhibiting signs of disease had virus present in their salivary glands more often than did the asymptomatic bats. In fact, in the latter group, rabies virus was recovered more often from brown fat than from salivary glands. Table IV shows in more detail the distribution of virus in the 35 infected, asymptomatic bats. 21 bats had virus in brain tissue alone and in 6 animals brown fat was the only tissue from which rabies virus was recovered. Four bats had virus present in brain and salivary glands, three had virus in brain and brown fat, and in one bat sacrificed 22 days after inoculation rabies virus was recovered from all three of the tissues tested.

The reports concerning the role of brown adipose tissue in the pathogenesis of rabies virus in insectivorous bats prompted other investigators to attempt isolation of the virus from the brown fat of naturally infected bats. In 1960, Bell and Moore reported the isolation of rabies virus from the pooled brown fat of two *Myotis* and from the brown fat of a single big brown bat *(E. fuscus)*. Dr. Bell was kind enough to send us this strain for use in

Table IV. Demonstration of rabies virus (canine strain) in tissues of experimentally infected *Tadarida brasiliensis* with no overt signs of disease[1]

Number of bats	Time after inoculation, days	Rabies virus demonstrated in		
		brain	salivary glands	brown fat
21	15–83	+	—	—
6	13–75	—	—	+
4	21–51	+	+	—
3	20–25	+	—	+
1	22	+	+	+
35	13–83	29	5	10

1 From Sulkin and Allen [1970].

additional studies on experimental rabies virus infection in bats. Not only did this bat strain prove to be more infective for *Myotis* than the canine strain used in our initial studies but the tissue tropism of this strain also differed significantly from that observed with the canine strain [Sulkin *et al.,* 1960]. 90% of the experimentally infected bats had virus present in interscapular brown adipose tissue, whereas virus was present in the brain tissue of only 50% of the infected animals. Thus, this rabies virus strain isolated from the brown fat of naturally infected bats exhibited marked lipotropic characteristics in the experimentally infected animals. Of even greater interest was the observation that infected animals showed no evidence of central nervous system disease. This strain of rabies virus produced a completely silent infection even though the virus multiplied in brain tissue as well as in the brown fat and salivary glands of bats.

Because the body temperatures of resting bats of certain species had been shown to parallel that of their environment over a wide range [Hock, 1951; Morrison, 1959] and because temperature was known to have an effect on the course and outcome of many experimental viral infections, our studies on experimental rabies virus infection in bats included controlled temperature experiments. Sadler and Enright [1959], controlling metabolic rate by holding bats at different ambient temperatures, had shown that following intracerebral inoculation of a highly neurotropic strain of bat rabies virus

into pallid bats, virus multiplied slowly in the tissues of animals held at
4 °C even though the infection was established by maintaining the animals
at 30 °C for a period of time before transferring them to a hibernating environment. In a continuation of studies of rabies virus infection in Mexican free-
tailed bats and little brown bats, variations in the environmental temperatures at which inoculated bats were maintained were introduced into the
experimental design [SULKIN et al., 1960]. In addition to the canine strain
of rabies virus used in the studies described above, the strain of rabies virus
isolated by BELL and MOORE [1960] from the brown adipose tissue of
naturally infected bats was used in the temperature studies. It was found
that both strains of virus failed to multiply in the tissues of bats placed
at 5 °C immediately after inoculation; however, virus survived in the cold-
exposed animals for many weeks, and upon transfer to a warmer environment the course of experimental rabies virus infection in these animals was
much the same as that observed in bats maintained at room temperature
from the time of inoculation [SULKIN et al., 1960]. These results suggested that
rabies virus could persist in naturally infected hibernating species of bats
during the winter months, and possibly undergo increased rates of multiplication upon arousal in the spring. Rabies virus was subsequently isolated
from the brain of an *Eptesicus* collected in New England in December, 1963
[GIRARD et al., 1965] and from the brain, brown fat and salivary glands of
another big brown bat collected in Montana in December, 1965 [BELL et al.,
1966].

Since the first insectivorous bat found to be infected with rabies virus in
the United States was a lactating female [SCATTERDAY, 1954] and in view
of the numerous reports indicating that the gravid state, a period of stress
accompanied by hormonal alterations, may affect the course of a viral infection, we undertook studies to determine the influence of the gravid state on
the course of experimental rabies virus infection in Mexican free-tailed bats
[SIMS et al., 1963]. Although we found no evidence that pregnancy increased
the susceptibility of *Tadarida* to the canine strain of rabies virus, evidence of
transplacental transmission of virus was obtained during the course of these
experiments. Rabies virus was demonstrated in the fetus of a bat shown to
be experimentally infected 23 days postinoculation. Virus was also isolated
from the brains of 3 newborn bats delivered 8–12 days after inoculation of
the mother bats; no virus was detected in any of the tissues from the mother
bats sacrificed at this time. Rabies infection in fetal tissues without concur-
rent infection in mother bats has been noted in studies with vampire bats in
Mexico [BAER, 1972]. Evidence suggesting transplacental transmission of

rabies virus in experimentally infected bats was also obtained in studies by CONSTANTINE et al. [1968]. If this event occurs even infrequently in naturally infected bats it could be an important factor in the persistence of rabies virus infection in bat populations in nature.

An interesting observation in the studies with gravid bats was that although susceptibility to experimental rabies virus infection was not altered significantly, frequency of virus isolation from the brown adipose tissue varied among groups of bats which received virus at different stages in the reproductive cycle. In groups of bats which received virus during the mid-gestation period and in the early *postpartum* period, virus was subsequently demonstrated in the brown fat of infected animals with equal frequency – 21.7 and 17.6%, respectively. In contrast, rabies virus was isolated from the brown fat of only 2.3% of the infected bats which received virus late in the gestation period. During the course of these studies, differences were noted in the gross appearance of the brown fat lobes of bats in various stages of the gestation period, and the amount of tissue present in animals at the time of delivery and in the early *postpartum* period was noticeably depleted. In view of these findings, further studies were undertaken to examine histologically the brown adipose tissue of the Mexican free-tailed bat throughout the reproductive cycle [SIMS et al., 1962]. In addition, since cortisone had been shown to greatly influence the brown fat of hamsters and mice [ARONSON et al., 1954] and of rats [LACHANCE and PAGE, 1953], the response of the brown fat of male bats to cortisone treatment was determined. These studies revealed marked variations in the lipid content of the brown fat of bats throughout the gestation period. Histological analyses showed that the amount of lipid in the brown fat cells increased as pregnancy progressed, reaching a peak in the later period, decreasing precipitously during and shortly after parturition.

Cortisone was found to exert a similar effect on the brown fat of male bats. Two days following treatment of bats with cortisone, the weight of the brown fat lobes was noticeably increased in comparison to the control groups. By the 5th day the weight of brown fat in the treated group was twice that of the controls, and 24 h later had begun to decline sharply, approaching the weight level of the control group. These results suggested that the lipid content of the interscapular brown adipose tissue of the Mexican free-tailed bat is highly mobile and subject to profound fluctuation during the gestation period as well as in response to cortisone treatment, and that adrenal activity plays an important role in directing alterations in the lipid content of the brown fat of bats. Having obtained information indicating that fluctuations in the lipid content of brown fat during the reproductive cycle influenced the multiplication of rabies virus in this tissue, experiments were designed to quantitate the effect of lipid deposition on the growth of rabies virus in brown fat cells. Since rabies virus inoculated intramuscularly into hamsters invaded and multiplied in the interscapular brown fat of these animals in a manner similar to that observed in bats [SULKIN et al., 1959] and since the lipid content of hamster brown adipose tissue could be increased by cortisone treatment [ARONSON et al., 1954], these experiments were carried out using the golden Syrian hamster [SIMS et al., 1967]. It was found that the growth of rabies virus in the brown adipose tissue of experimentally infected hamsters was inhibited by an increase in the lipid content of brown fat cells induced by the administration of cortisone. However, despite the low quantities of rabies virus demonstrable in the brown fat of rabies-infected hamsters which had received cortisone, marked evidence of histopathology was observed in the tissue. In contrast, brown fat tissue from infected hamsters which were not treated with cortisone contained high titers of rabies virus in the absence of degenerative alterations. The demonstration of areas of hemorrhage and necrosis in rabies-infected brown fat of cortisone-treated hamsters could be correlated with isolation of rabies virus from the blood of animals sacrificed when central nervous system symptoms were evident, suggesting that the viremia was due to spill-over of virus from brown adipose tissue. Assuming a similar situation occurs in infected pregnant bats in nature, a focus of rabies virus infection in the brown fat would increase the chances for viremia and possible transplacental transmission of rabies virus in these animals.

In subsequent studies by other investigators on experimental rabies virus infection in various bat species, virus strains isolated from naturally infected bats of the same species were used and overt signs of central nervous system disease and/or death were observed [CONSTANTINE, 1966a, b; CONSTANTINE and WOODALL, 1966; CONSTANTINE et al., 1968]. Infection was generally widespread, being demonstrable in brain, brown fat, kidney and lung tissue, and in salivary glands and saliva. Bite transmission of rabies virus from bats showing signs of irritability and aggressiveness to foxes, coyotes, mice and hamsters was accomplished. Although no evidence of inapparent rabies virus infection in bats was obtained in these studies, recovery from overt infection was demonstrated in one experiment in which 5 bats survived an observation period of 108 days and had returned to normal health when sacrificed; rabies virus could not be demonstrated in any tissues taken from these animals [CONSTANTINE, 1966b]. BAER and BALES [1968] studied rabies virus infection in Mexican free-tailed bats inoculated with a strain of virus isolated from pooled salivary gland tissue of naturally infected *Tadarida*. Infection rates varied according to dose and route of inoculation and dissemination of virus in infected bats was most widespread in animals inoculated intracerebrally, being demonstrable in brain, brown fat, salivary gland, kidney and lung tissue. All infected animals developed signs of rabies and died with the exception of a single bat which was sacrificed in a seemingly healthy condition 181 days following intramuscular inoculation; rabies virus was isolated from the brain, brown fat, salivary glands, kidneys and lungs of this animal.

Growth of Rabies Virus in Bat Tissues in vitro

Another approach to gaining a better understanding of the ability of bats to sustain inapparent rabies virus infection was the investigation of the growth of rabies virus in bat cells in tissue culture preparations; this would allow the application of more definitive techniques to the study of cell-virus interactions. Because of the proposed importance of the growth of rabies virus in the brown fat of experimentally and naturally infected bats [SULKIN, 1962], cultivation of this tissue constituted the first step in the *in vitro* studies [ALLEN *et al.,* 1964a]. Preliminary efforts to prepare tissue cultures of bat brown fat met with variable success which could be correlated with the season in which the animals were netted for these studies. Investigators studying the nature of brown adipose tissue had found, particularly in hibernating species, that the biochemical constituents of the tissue vary with the season and that the fluctuations in tissue components were reflected grossly in the color and weight of the interscapular brown fat lobes [JOHANNSON, 1959]. Therefore, in an attempt to carry out *in vitro* studies on a year-round basis, data on the wet weight, coloration and cultivability of adult bat brown fat were collected at different seasons to determine optimum growth periods. Observations were made on 3 bat species: deep hibernators, *M. lucifugus* and *E. fuscus,* and the quasi-hibernator, *T. brasiliensis*. Growth of brown adipose tissue in explant and monolayer cultures was accomplished with greatest success when obtained from *Myotis* or *Eptesicus* netted in the fall, just prior to hibernation, or from pregnant *Tadarida* collected in the spring, the seasons when these species appear in peak physical condition and when the brown fat lobes reach maximum weight and are light brown in color. Attempts to culture the brown fat of bats taken from hibernation or sacrificed during or shortly after parturition when the tissue was markedly less abundant and a dark red color were not consistently fruitful. These observations, indicating that the degree of success achieved in culturing brown fat could be correlated with seasonal variations in the physiological activity of this tissue, suggested that the major portion of the tissue culture populations was derived from brown fat cells rather than from a minority cell type present in the tissue. In the initial studies on growth of rabies virus in cultured bat brown fat cells, primary or first passage cultures were used with the hope that the brown fat cells would retain characteristics of the whole tissue in the intact host. In this regard, it was demonstrated that insulin added to the culture medium stimulated the synthesis of glycogen in monolayer cultures of bat brown adipose tissue in a manner similar to that reported by SIDMAN [1956] for organ cultures of rat brown fat and analogous to the action of insulin on brown adipose tissue of the intact animal [WERTHEIMER, 1945].

In general, cultured bat brown adipose tissue cells are fibroblast-like in appearance, and primary and first passage cultures can be maintained for prolonged periods of time at 37 °C by changing culture media at 4- to 5-day intervals. Monolayer cultures on coverslips in Leighton tubes were shown to support the growth of a bat strain of rabies virus for at least 56 days with no gross degenerative changes observed [ALLEN et al., 1964a]. Virus could be detected in culture fluids and in cells by mouse inoculation, but the direct fluorescent antibody technique (FAT) as described by GOLDWASSER et al. [1959] proved to be a much more sensitive method for demonstrating cell-associated virus. Infected cultures contained cells with diffuse fluorescence and cells with areas of more intense fluorescence, appearing in sharply outlined aggregations of various sizes and shapes in the cytoplasm. When stained by a modified May-Grunwald-Giemsa procedure (MGG), the aggregations which fluoresced specifically as rabies virus antigen stained as blue inclusion bodies surrounded by halos. These intracytoplasmic inclusion bodies appeared similar to those observed by other investigators in earlier studies on the growth of rabies virus in a variety of tissue culture systems [KISSLING, 1958; ATANASIU and LEPINE, 1959; KISSLING and REESE, 1963; FERNANDES et al., 1963].

In the preliminary studies on the characterization of cultured bat brown fat it was noted that cultures of this tissue could be maintained for extended periods of time at 8 °C. Thus, a study of the persistence of rabies virus in this tissue culture system at low temperature was carried out to obtain information which could be related to the possible persistence of rabies virus infection in hibernating bats in nature [ALLEN et al., 1964b]. Cultures of bat brown fat retained rabies virus infection for more than 4 months at 8 °C, supporting the hypothesis that rabies virus particles present in the inter-scapular brown adipose tissue of the intact bat would remain viable through-out periods of hibernation. Conclusive evidence of virus multiplication at 8 °C was not obtained, although there was the suggestion that some level of viral activity was maintained in the cold since transfer of cultures from 8 to 37 °C resulted in rapid activation of rabies virus growth cycles.

Since rabies virus had been shown to invade and multiply in the inter-scapular brown fat of experimentally infected hamsters [SULKIN et al., 1959] in the absence of histopathological alterations in the tissue [SIMS et al., 1967], the growth of a bat strain of rabies virus in cultured hamster brown fat was studied. Characteristics of rabies virus infection in monolayer cultures of hamster brown fat were similar to that observed in bat brown fat cultures; a persistent infection was established with no evidence of a cytopathic effect other than the formation of intracytoplasmic inclusion bodies which did not appear to interfere with cellular function. A cytochemical study of rabies virus-induced inclusion bodies in hamster brown fat cells revealed the pres-ence of ribonucleic acid, protein and certain protein-bound groups within the inclusions [LEONARD et al., 1967].

Monolayer cultures of bat kidney cells were also shown to be susceptible to infection with the bat strain of rabies virus used in the studies with cultured bat brown fat. Unlike the persistent, endosymbiotic infection ex-hibited by this strain of rabies virus in bat brown fat cultures, bat kidney cultures were destroyed by the growth of the virus [SIMS, 1969]. The cyto-pathic effect observed was characterized by a balling-up and ballooning of the cells followed by lysis. Another indication of the increased susceptibility of cultured bat kidney cells to rabies virus was the observation that a canine strain of the virus, of low infectivity for cultured bat brown fat cells, could be propagated in bat kidney cells. Rabies virus has been isolated from the kidneys of naturally and experimentally infected bats, and the presence of rabies virus in bat urine as a factor in the aerosol transmission of rabies in bat caves has been discussed [CONSTANTINE, 1966; CONSTANTINE and WOOD-ALL, 1966; CONSTANTINE, 1967].

Because the brown adipose tissue from deep-hibernating species of bats could be cultured most consistently when obtained from animals netted in the fall just prior to hibernation and because pregnant bats for embryo cultures could only be obtained in the spring of the year, our initial in vitro studies were limited to these seasons. In order to alleviate these restrictions and have an unlimited supply of bat cell cultures for use in studies on rabies virus and other viruses associated with bats, efforts were made to establish stable lines of bat cells which retained 'bat' characteristics. Our first success was achieved with the development and characterization of cell lines derived from embryos of Mexican free-tailed bats and designated *T. brasiliensis* embryo (TBE) cells. Morphologically different cell lines prepared from these embryos, as well as cloned lines derived from them, are diploid (2n = 48), and remained so for at least 15 months (45 passages), indicating unusual stability [STEWART *et al.,* 1970]. Primary and first passage cultures of TBE cells sustained persistent infection with rabies virus in a manner similar to that observed with bat brown fat cells, and the cell lines and cloned lines derived from them retained this characteristic [SIMS, 1969]. It has also been possible to establish a line of bat brown fat (BBF) cells which can be maintained by serial passage in the laboratory and which appear to have retained characteristics recorded for the primary and first passage cultures of this tissue used in earlier studies. BBF cells chronically infected with rabies virus have been carried through numerous passages over a period of more than 1 year. Detailed studies on cell-virus interactions in these cultures have shown that rabies virus infected cells bearing intracytoplasmic inclusion bodies grow at the same rate as uninfected BBF cells [M. MANTANI, R. ALLEN and S. E. SULKIN, unpublished observations].

Arboviruses

Epidemiological Considerations

The arboviruses present perhaps the most complex biological life cycles of any of the known viral agents. Simply stated, an infection chain depends on the association of a viremic host, a blood-sucking vector and a susceptible host; however, the multiplicity of vectors and hosts which may be involved in the natural history of any one virus is extremely difficult to determine. Since the early studies on yellow fever by REED [1911], in which it was first shown that transmission of a virus infection of man was dependent on an arthropod vector, the ecology of members of this group of viruses has been the subject of many studies. The arboviruses associated with illness in man and domestic animals have received special attention, and the identification of specific vectors and wild-life hosts of these agents has been an invaluable aid in establishing methods for the control of these diseases. When the eradication of the arthropod vector of yellow fever virus failed to prevent recurrent epidemics of this disease, the existence of a monkey-mosquito transmission cycle was discovered [STRODE, 1951]. Jungle or sylvan yellow fever provides one of the earliest and perhaps best examples of the importance of an animal reservoir in the perpetual existence of a viral agent of human disease. Whereas monkeys are perhaps the most important known reservoir hosts of yellow fever virus, early epidemiological studies on Eastern equine encephalomyelitis (EEE) virus [GILTNER and SHAHAN, 1933; TEN-BROECK et al., 1935] led to the designation of birds as the most frequently recognized natural hosts of the viruses of the North American encephalitides and of Japanese B encephalitis (JBE) virus [HAMMON, 1958; SCHERER and BUESCHER, 1959; KISSLING, 1960].

In tropical areas arbovirus transmission cycles may be continuous throughout the year, providing constant foci of infection as sources for recurrent epidemics in man and domestic animals. Although the basic transmission cycles which operate to keep arboviruses circulating in nature during the spring, summer and fall months in endemic areas have been defined, little is known about how these viruses persist in nature during the winter months.

It has been considered that the viruses do not remain in apparent endemic areas throughout the winter but are reintroduced annually by migratory reservoir hosts. Many studies have been concerned with the movement of EEE and Western equine encephalomyelitis (WEE) viruses by birds. Initial attempts failed to isolate virus from birds migrating into this country in the spring [KISSLING et al., 1957], but recently EEE virus isolates, serologically identifiable as South American strains, were obtained from birds captured in south Louisiana apparently during their northward migration from South America [CALISHER et al., 1971]. There is also evidence that migratory birds may transport EEE and WEE viruses southward from the United States in the fall [STAMM and NEWMAN, 1963; LORD and CALISHER, 1971], and the possibility remains that migrating bird species of Japan may influence the ecology of JBE virus in that country [SCHERER et al., 1959]. Whether or not these transported viruses actually become established in the new locales is difficult to determine, however, and it seems more likely that the mosquito-borne arboviruses of temperate zones are truly endemic and are maintained in nature in chronically infected hosts [MATUMOTO, 1969; LORD, 1970].

Investigations to provide supportive evidence that the mosquito vectors themselves harbor the agents in a latent state during the winter have been concerned with field studies designed to recover virus from mosquitoes captured during the winter months [BLACKMORE and WINN, 1956; REEVES et al., 1958a] and with attempts to demonstrate virus transmission by experimentally infected mosquitoes following prolonged periods at low temperatures [HURLBUT, 1950; BELLAMY et al., 1958]. Sufficient positive results were not obtained in any of these studies, however, to permit firm conclusions concerning the magnitude of the role of mosquitoes in the overwintering of the arboviruses. Experimental cyclic transmission of WEE virus in chickens and Culex tarsalis through a 1-year period has been demonstrated but the authors concluded that, although this virus persisted in its vector through the winter period under experimental conditions, the chances of this occurring in nature were small [BELLAMY et al., 1967]. The current opinion holds that this is probably not a mechanism which contributes to the persistence of these agents in nature [CHAMBERLAIN, 1968].

The ability of vertebrate hosts to support latent arbovirus infections during periods when the mosquito vectors are inactive and to exhibit recurrent viremias would provide excellent means for the perpetuation of these viruses in nature. Experimental support for such a theory has been obtained with a variety of vertebrate hosts and it has been suggested that wild birds could serve to maintain arboviruses in nature either by maintaining chronic latent infections throughout the winter with recurrent viremias in the spring or by low-level mosquito-bird-mosquito transmission cycles during the winter months [REEVES, 1961]. In studies of experimental infection of wild birds with WEE virus, REEVES et al. [1958 b] demonstrated persistence of virus in various organs for prolonged periods of time as well as recurrent viremia several months after inoculation. In an experimental study with wild birds, St. Louis encephalitis (SLE) virus was isolated from the gizzard of a cowbird 38 days after infection [CHAMBERLAIN et al., 1957]. However, in rather extensive surveys of birds wintering in the southern United States, either as permanent residents or as migrants, no virus isolates were obtained [KISSLING et al., 1955, 1957].

Certain poikilothermic animals have been studied as possible overwintering hosts for a number of arboviruses. Studies on JBE and WEE virus infection in frogs [CHANG, 1958; BURTON et al., 1966] and EEE virus infection in lizards [KARSTAD, 1961] have indicated that these cold-blooded animals could maintain these viruses through the winter months. The most extensively investigated poikilotherm has been the snake; there is considerable evidence from both experimental and field studies to indicate that it could play an important role in the persistence of WEE virus in nature [BURTON et al., 1966; GEBHARDT and STANTON, 1966; GEBHARDT et al., 1973].

The endemicity of arboviruses is proof that some mechanism which provides for their persistence in nature does operate. Since at least some supportive evidence has been obtained for each of the theories proposed to explain overwintering of arboviruses, it seems likely that no single mechanism is sufficient to account for the phenomenon, but rather a combination of events is required to insure the continued existence of these agents. Thus, the search continues for yet unknown animal reservoirs and arthropod vectors in an effort to explain the persistence of these agents despite interruptions in the natural transmission cycles. There is now a significant amount of evidence indicating that bats may be reservoir hosts particularly suited to the year-round maintenance of arboviruses in nature. Tables V and VI present evidence, cited from the literature or obtained through personal communications with investigators, of natural and experimental infections of bats with arboviruses. In many instances the field studies were either initially prompted by or subsequently supported by the results of studies on experimental infections of bats with a particular virus. We have tabulated all the evidence suggestive of natural arbovirus infections in bats which has come to our attention; in many instances, particularly those based on serological evidence alone, firm proof that the bat-virus association exists in

Table V. Natural arbovirus infections in bats

Virus	Bat genera	Geographic area	Evidence of infection	References
Group B				
Japanese B encephalitis	*Pteropus, Hipposideros, Miniopterus[1], Cynopterus, Myotis, Rhinolophus, Murina, Pipistrellus, Plecotus, Vespertilio*	Australia, Taiwan, Thailand, India, Malaysia, Japan	virus isolation (Bl, BF, Ki)[2] N, HI antibodies	ROWAN and O'CONNOR, 1957; WANG *et al.*, 1962; RUSSELL, 1968; CAREY *et al.*, 1968; cited by CONSTANTINE, 1970; SULKIN *et al.*, 1970; MIURA *et al.*, 1970; CROSS *et al.*, 1971.
St. Louis encephalitis	*Tadarida, Eptesicus, Rhinophylla*	United States, Brazil	virus isolation (Bl, Sp) N, HI antibodies	WHITNEY, 1963; SULKIN *et al.*, 1966; RUEGER *et al.*, 1966; cited by CONSTANTINE, 1970; ALLEN *et al.*, 1970.
Rio Bravo	*Tadarida, Eptesicus*	United States	virus isolation (SG, Bl, BF, Lu, Pa)	BURNS *et al.*, 1956; JOHNSON 1962; CONSTANTINE and WOODALL, 1964; SULKIN *et al.*, 1966; BAER and WOODALL, 1966; ALLEN *et al.*, 1970.
Entebbe bat	*Tadarida, Eidolon, Rousettus, Rhinolophus*	East Africa	virus isolation (SG) HI antibodies	LUMSDEN *et al.*, 1961; SHEPHERD and WILLIAMS, 1964; SIMPSON *et al.*, 1968.
Dakar bat	*Scotophilus, Otomops, Tadarida*	Senegal, East Africa, Nigeria	virus isolation (SG, Sa, Pa) HI antibodies	WILLIAMS *et al.*, 1964; BRÉS and CHAMBON, 1964; CRAIG and SSENKUBUGE, 1968; HENDERSON *et al.*, 1968; SIMPSON *et al.*, 1968; KARABATSOS, 1969.

Disease	Bat genera	Location	Detection	References
Montana *Myotis* leucoencephalitis	*Myotis*	United States	virus isolation (SG, Sa, Br, BF)	BELL and THOMAS, 1964.
Bukalasa bat	*Tadarida*	East Africa	virus isolation (SG), HI antibodies	WILLIAMS et al., 1964; FUKUMI et al., 1967.
Bangui bat	unidentified	Central Africa	virus isolation (SG)	CHIPPAUX and CHIPPAUX-HYPPOLITE, 1965.
Mount Suswa bat	*Otomops*	East Africa	virus isolation (SG)	HENDERSON et al., 1968.
Yellow fever	*Glossophaga, Eidolon, Tadarida, Epomophorus, Rousettus, Eptesicus*	Brazil, Colombia, East Africa, Ethiopia	virus isolation (Br), N, HI antibodies	LAEMMERT et al., 1946; SHEPHERD and WILLIAMS, 1964; WILLIAMS et al., 1964; ANDRAL et al., 1968; SERIE et al., 1968; SIMPSON et al., 1968.
Tickborne encephalitis	*Barbastella, Myotis, Plecotus, Rhinolophus*	Central Bohemia, Italy	N, HI antibodies	HAVLIK and KOLMAN, 1957; VERANI, 1968.
Kyasanur Forest disease	*Cynopterus, Pteropus, Rousettus, Rhinolophus*[3]	India	virus isolation (Sp), N, HI antibodies	PAVRI and SINGH, 1965, 1968; RAJAGOPALAN et al., 1969.
West Nile	*Pteropus, Eidolon, Tadarida, Rousettus, Rhinolophus,* unidentified	Egypt, Australia, East Africa, Israel, Ethiopia, India, Italy	virus isolation (Sp), N, HI antibodies	TAYLOR et al., 1956; ROWAN and O'CONNOR, 1957; SHEPHERD and WILLIAMS, 1964; AKOV and GOLDWASSER, 1966; BALDUCCI, 1968; SIMPSON et al., 1968; ANDRAL et al., 1968; PAUL et al., 1970.
Zika	*Tadarida, Eidolon, Rousettus,* unidentified	East Africa, Ethiopia	HI antibodies	SHEPHERD and WILLIAMS, 1964; SIMPSON et al., 1968; ANDRAL et al., 1968.

Table V (continuation)

Virus	Bat genera	Geographic area	Evidence of infection	References
Dengue 1, 2 and/or 4	*Pteropus, Cynopterus, Eonycteris, Hipposideros, Scotophilus*	Australia, India, Malaysia	N, HI antibodies	ROWAN and O'CONNOR, 1957; SHAH and DANIEL, 1966; CAREY et al., 1968; cited by CONSTANTINE, 1970.
Ntaya	*Eidolon*	East Africa	HI antibodies	SIMPSON et al., 1968.
Usutu	*Eidolon, Rousettus*	East Africa	HI antibodies	SIMPSON et al., 1968.
Ilheus	*Carollia, Glossophaga,* unidentified	Colombia, Brazil	N, HI antibodies	PRIAS-LANDINEZ, 1966; cited by CONSTANTINE, 1970.
Murray Valley encephalitis	*Eptesicus, Pteropus*	Australia	N, HI antibodies	ROWAN and O'CONNOR, 1957; STANLEY and CHOO, 1964.
Tembusu	*Pteropus*	Malaysia	N antibodies	cited by CONSTANTINE, 1970.
Wesselsbron	*Cynopterus*	Thailand	N antibodies	RUSSELL, 1968.
Bussuguara	unidentified	Brazil	HI antibodies	cited by CONSTANTINE, 1970.
Uganda S	*Eidolon, Eptesicus, Tadarida, Miniopterus,* unidentified	East Africa, Ethiopia	HI antibodies	SHEPHERD and WILLIAMS, 1964; ANDRAL et al., 1968.
Turkey meningo-encephalitis	*Rousettus*	Israel	HI antibodies	AKOV and GOLDWASSER, 1966.
Uncharacterized	*Miniopterus, Myotis, Pteropus, Cynopterus*	Japan, India, Cambodia	virus isolation (BI, SG, BF) HI antibodies	CAREY et al., 1968; SALAUN, 1970; SULKIN et al., 1970.
Group A				
Chikungunya	*Eidolon, Hipposideros, Scotophilus, Tadarida*	East Africa, India, Senegal	virus isolation (SG) HI antibodies	SHEPHERD and WILLIAMS, 1964; BRÉS and CHAMBON, 1964; BRÉS et al., 1964.

Virus	Bat genera	Location	Finding	Reference
Semliki Forest	*Tadarida*	East Africa	HI antibodies	SHEPHERD and WILLIAMS, 1964.
Mayora	*Carollia*	Colombia	HI antibodies	PRIAS-LANDINEZ, 1966.
Mucambo	*Carollia, Artibeus*	Brazil	N, HI antibodies	cited by CONSTANTINE, 1970.
Pixuna	unidentified	Brazil	HI antibodies	cited by CONSTANTINE, 1970.
Una	unidentified	Brazil	HI antibodies	cited by CONSTANTINE, 1970.
Western equine encephalitis	*Eptesicus*, unidentified	United States, Brazil	virus isolation (SG) HI antibodies	GOLDFIELD, 1968; cited by CONSTANTINE, 1970.
Eastern equine encephalitis	*Eptesicus, Myotis, Tadarida, Lasiurus, Artibeus*	United States, Brazil	virus isolation (Br, SG, Li-Sp pools) N, HI antibodies	KARSTAD and HANSON, 1958; DANIELS *et al.*, 1960; MAIN, 1970; GOLDFIELD, 1970; cited by CONSTANTINE, 1970.
Venezuelan equine encephalitis	*Carollia, Artibeus, Glossophaga, Desmodus*	Colombia, Panama, Mexico, Brazil	virus isolation (Bl, He-Li-Sp pools, He-Lu-Sp pools) N, HI antibodies	PRIAS-LANDINEZ, 1966; GRAYSON and GALINDO, 1968; SCHERER *et al.*, 1971; CORREA-GIRON *et al.*, 1972; cited by CONSTANTINE, 1970.

1 JBE virus isolated from mosquitoes (*Culex annulus*) collected on same day from same cave as virus-infected *Miniopterus* [CROSS *et al.*, 1971].

2 Tissue abbreviations: Bl = blood; Br = brain; BF = brown fat; He = Heart; Ki = kidney; Li = liver; Lu = lung; Pa = parotid gland; Sa = saliva; SG = salivary gland; Sp = spleen.

3 KFD virus isolated from adult ticks (*Ornithodoros* sp.) collected from same abandoned well inhabited by virus-infected *Rhinolophus* [RAJAGOPALAN *et al.*, 1969].

Table VI. Experimental arbovirus infections in bats

Virus	Bat genera	Characteristics[1] of infection	References
Group B			
Japanese B encephalitis	*Pipistrellus, Eptesicus, Myotis, Tadarida*	Bl[2], Br, BF, Ki; N antibodies; recurrent viremia; transplacental transmission; inapparent	Ito and Saito, 1952; Corristan et al., 1956; LaMotte, 1958; Sulkin et al., 1960a, 1963, 1964, 1966b, c.
St. Louis encephalitis	*Myotis, Tadarida*	Bl, Br, BF, Ki; transplacental transmission; inapparent	Sulkin et al., 1960a, 1963, 1964, 1966b.
Rio Bravo	*Tadarida, Eptesicus*	Bl, Br, BF, SG, Sp; inapparent	Sulkin and Allen, unpublished observations.
Entebbe bat	*Rousettus*	Bl; inapparent	Simpson and O'Sullivan, 1968.
Yellow fever	*Epomophorus, Eptesicus, Eidolon, Rousettus, Tadarida*	Bl, Br, BF, Sp; inapparent	Rodhain, 1936; Sulkin, 1962; Simpson and O'Sullivan, 1968; Sulkin and Allen, unpublished observations.
Tickborne encephalitis	*Barbastella, Myotis, Plecotus*	Bl, Br, BF, He, Li, Sp; symptoms and pathology; N antibody	Kolman et al., 1960; Nosek et al., 1961.
Kyasanur Forest disease	*Rousettus, Cynopterus*	Bl, Br, SG, He, Lu, Ki, Li, Sp; symptoms; N, HI antibodies	Pavri and Singh, 1965, 1968.
West Nile	*Rousettus*	Bl; N antibodies	Simpson and O'Sullivan, 1968.
Zika	*Rousettus, Eidolon*	Bl	Shepherd and Williams, 1964; Simpson and O'Sullivan, 1968.
Ntaya	*Tadarida*	Bl	Simpson and O'Sullivan, 1968.
Usutu	*Rousettus*	Bl	Simpson and O'Sullivan, 1968.

Group A

Chikungunya	*Pteropus*	Bl; N, HI antibodies	SHAH and DANIEL, 1966.
Semliki Forest	*Eidolon, Rousettus, Tadarida*	Bl, Br, SG, Ki; inapparent	SIMPSON and O'SULLIVAN, 1968.
Sindbis	*Eidolon*	Bl; N antibodies	SIMPSON and O'SULLIVAN, 1968.
Western equine encephalitis	*Tadarida*	Bl[2]; inapparent	cited by CONSTANTINE, 1970.
Eastern equine encephalitis	*Eptesicus*	Bl; inapparent	LAMOTTE, 1958.
Venezuelan equine encephalitis	*Eptsicus, Myotis, Pipistrellus, Plecotus*	Bl[2], Bt, BF, Sp; N antibodies; inapparent	CORRISTAN *et al.*, 1956, 1958; SULKIN and ALLEN, unpublished observations.

1 Tissue abbreviations: Bl = blood; Br = brain; BF = brown fat; He = heart; Ki = kidney; Li = liver; Lu = lung; SG = salivary gland; Sp = spleen.
2 Mosquitoes infected by feeding on bats.

nature to a significant degree is lacking. Isolation of a virus from the tissues of naturally infected bats, and especially from the blood, provides stronger evidence that a relationship may exist, but even then, a single isolation from bats in a particular locale may have little meaning for that area. Repeated samplings of the bat population in a given locale involving a total of several hundred animals captured at intervals throughout the year should be accomplished before the importance of bats in that area as reservoir hosts for a particular agent can be evaluated. Much of the tabulated information was derived from studies carried out within the past 10 years, pointing up the current interest in the bat as a reservoir host for the arboviruses; all of the evidence presented, no matter how slight in some instances, indicates a need for the continued inclusion of these animals in epidemiological surveys. The world-wide distribution of the geographic areas reporting natural arbovirus infections in bats reflects the concomitant ubiquitousness of the arboviruses and this mammalian host.

Group B Arbovirus Infections in Bats

Susceptibility of Bats to Experimental Infection with JBE Virus

The first firm evidence of the susceptibility of bats to an arbovirus was that reported by ITO and SAITO [1952] who infected pipistrella bats *(Pipistrellus abramus)* with JBE virus by intracerebral inoculation. Virus was detected in brain tissue but not in blood, liver, pancreas or gastric contents, and could be serially passed in bats without loss of virulence. Although brain tissue titers ranged as high as 10^{-7} none of the infected bats showed signs or symptoms of encephalitis nor did they develop neutralizing antibodies. Experimental infection of two species of insectivorous bats *(E. fuscus* and *P. subflavus)* with JBE virus was reported by CORRISTAN *et al.* [1956]. Subcutaneous inoculation induced viremia which persisted for at least 15 days in some bats. The animals survived infection without overt signs of illness. LAMOTTE [1958] presented experimental evidence of mosquito-bat-mosquito transmission cycles with JBE virus at room temperature. Mosquitoes were also shown to feed on bats at simulated cold weather cave temperature (10 °C) and to transmit infection under these conditions. In one instance, a *P. subflavus* developed viremia 9 days following ingestion of 3 infected mosquitoes, suggesting that bats may become infected by feeding on mosquitoes in nature.

These reports, together with the field and experimental evidence of the effectiveness of insectivorous bats as reservoir hosts for rabies virus, prompted a more intensive investigation of the role of bats in the ecology of JBE virus. Studies initiated in our laboratory were aimed at learning more about the susceptibility of different bat species to strains of JBE virus and obtaining additional information concerning the duration of the viremic phase of the infection, the tissues which supported virus growth, the influence of temperature on the course of infection and the immune response in the experimentally infected bat [SULKIN et al., 1963, 1966a, b]. The purpose of these studies was to accumulate data which would aid in determining if large-scale surveys of bat populations in Japan for evidence of natural JBE virus infection were warranted and, if so, to provide information which would be useful in planning the field work. Three bat species (T. brasiliensis, M. lucifugus and E. fuscus) and two strains of JBE virus – the mouse-adapted Nakayama strain [KASAHARA et al., 1936] and an unadapted mosquito isolate designated OCT-541 [ROHITAYODHIN and HAMMON, 1962] – were used in the initial studies. In all experiments each animal received approximately 150 weanling mouse intracerebral LD_{50} ($WMICLD_{50}$) of virus subcutaneously. Although both strains of JBE virus were infective for Tadarida and Myotis, the intensities of the infections varied, being less widespread with the high mouse passage Nakayama strain. In experiments with Eptesicus only the OCT-541 strain of JBE virus, previously shown to be highly infective for this bat species, was used. The characteristics of JBE virus infection in bats approached what could be termed the ideal infectious process in a reservoir host, e.g. a long-term initial viremia which would favor mosquito transmission and the establishment of an asymptomatic, latent-type infection which would be conducive to recurrent viremias. Bats inoculated subcutaneously with small doses of virus developed viremia within 24–72 hours; infection rates were virtually 100 % and many animals circulated virus for as long as 25–30 days at levels well above those necessary to infect mosquitoes.

The importance of an animal host in the biological life-cycle of an arbovirus depends on the establishment of a viremic state of variable duration which provides infective virus for feeding vectors. Whether or not replication of arboviruses actually occurs in the blood of susceptible animals has never been conclusively determined, but it is known that various host tissues other than blood are involved in the infectious process. Recurrent viremia in a natural host in the absence of reinfection would depend on the presence of latent foci of infection in one or more tissues from which active virus could be shed periodically into the bloodstream. A major interest in these studies, therefore, was the determination of the tissue tropisms of JBE virus strains in the various bat species [SULKIN *et al.*, 1963]. Following subcutaneous inoculation, evidence of virus replication was demonstrated in brown adipose tissue of infected animals for extended periods of time and, to a lesser degree, in the brains and kidneys of some animals. Viral invasion and multiplication in the brown adipose tissue of these animals seemed particularly significant, since this tissue could sequester virus particles in an inactive but viable state during the dormant period and could then provide seed for a recurrent viremia upon arousal in the spring, thus enabling hibernating species to harbor certain arboviruses over the winter in temperate zones. Although virus was demonstrated in low titer in the brain tissue of a few bats, none showed overt signs of encephalitis and sections of infected brains exhibited no evidence of viral pathology. These results indicated that either the brain tissue of the bat is capable of supporting JBE virus multiplication without suffering injury, or that so few brain cells are infected following subcutaneous inoculation their destruction by viral multiplication is not sufficiently injurious to produce encephalitis in the bats nor to be detected by histopathological techniques. Subsequently, it was found that the intensity of JBE virus infection in bat brain tissue was influenced by the route of inoculation. In big brown bats *(E. fuscus)* inoculated intracerebrally, virus titers in brain and brown fat tissue reached concentrations approaching the levels demonstrable in brains of mice dying of infection, yet the bats appeared to suffer no ill effects and evidence of viral pathology in these tissues was minimal or absent. JBE virus was also demonstrated in high titer in the blood and kidneys of intracerebrally inoculated *E. fuscus*. This was at variance with the report by ITO and SAITO [1952], who observed that a strain of JBE virus similarly inoculated into pipistrellas *(P. abramus)* multiplied to high titer in brain tissue but was not demonstrable in blood or visceral organs. This variance was probably due to differences in virus strains and bat species, since our studies indicated that JBE virus infection following intracerebral inoculation in *Myotis* was less intense and widespread than in *Eptesicus* [SULKIN *et al.*, 1963].

Transplacental Transmission of JBE Virus in Experimentally Infected Bats

An extensive literature documents evidence of transplacental transmission of many viruses in both man and animals [EBERT and WILT, 1960; BLATTNER

Table VII. Transplacental transmission of JBE (OCT-541 strain) virus in experimentally infected *Tadarida brasiliensis*[1]

Stage of pregnancy	Days after inoculation of gravid bats	Transplacental transmission
Early	5–10	3/15[2]
Mid	5–9	2/11
Late	7–18	9/11

1 Data taken from SULKIN *et al.* [1964].
2 Number of infected fetuses per number of viremic gravid bats.

and HEYS, 1961], and a few reports concern the arboviruses [MEDOVY, 1943; BURNS, 1950; SHINEFIELD and TOWNSEND, 1953; SHIMIZU *et al.,* 1954]. BURNS [1950] found that JBE virus was passed from naturally infected mother sows to their fetuses *in utero,* resulting in abortion and stillbirth; similar findings were obtained in pregnant sows experimentally infected with this virus [SHIMIZU *et al.,* 1954]. Since information concerning transplacental transmission of JBE virus in a natural host other than swine was lacking, this phenomenon was studied in Mexican free-tailed bats experimentally infected with JBE virus while in progressive stages of pregnancy [SULKIN *et al.,* 1964]. Although JBE virus was found to cross the placenta and invade fetuses of gravid bats infected during early, mid and late stages of the gestation period, transplacental passage of the agent occurred most frequently during the late gestation period (table VII). Distribution of JBE virus in fetal tissues paralleled that in infected mother bats; virus was found in fetal brown fat, brain, kidneys and carcass. All fetuses were viable and no gross pathology was noted. These results suggested that JBE virus could be perpetuated in nature in bat populations by direct passage from infected mothers to their offspring.

Influence of Environmental Temperature on Experimental JBE Virus Infections in Bats

In early studies by ITO and SAITO [1952] it was shown that multiplication of JBE virus in bats was influenced by environmental temperature, being depressed in hibernating bats and most active in bats held at 30 °C. LAMOTTE

[1958] also observed that bats experimentally infected with JBE virus and placed at 10 °C immediately after inoculation sustained the infection for several months. As mentioned earlier, experimental rabies infection also persists in bats maintained for long periods at low temperature [SADLER and ENRIGHT, 1959; SULKIN et al., 1960]. Since the physiological functions necessary for survival of bats continue at a reduced rate at the low temperatures experienced during hibernation, it is not surprising that the tissues of such animals sustain virus particles in a latent state or perhaps even provide adequate conditions for a slow rate of replication. Whereas bats in temperate zones experience long periods at low temperature during the winter months, they likewise are subject to periods at high temperature during the summer months. Species which are commonly found roosting in attics during the daylight hours in the summertime experience prolonged periods of high body temperature, and night flights in search of food also cause an increase in body temperature. More extensive studies to determine the degree to which periods of exposure to low (5–10 °C) and high (37 °C) temperatures would influence initiation and persistence of experimental arbovirus infection in bats seemed indicated before reasonable speculations could be made on how seasonal variations in environmental temperature would affect the manner in which these animals could serve as year-round hosts for certain arboviruses. When such studies were initiated in our laboratory, it was noted that bats netted in the fall of the year just prior to entrance into hibernation in nature survived longer in a cold environment in the laboratory than did those obtained during the spring or summer months. This was not surprising since these animals are known to prepare for hibernation in late summer and fall by storing fat and liver glycogen [TROYER, 1959; DAVIS, 1970]. It has also been shown that certain physiological alterations which deal with the hibernating bat's ability to arouse spontaneously in the cold occur in the fall [MENAKER, 1962]. In addition, the inability of the summer bat to raise its body temperature in a cold environment has been demonstrated [MENAKER, 1962], suggesting that hypothermia in bats in the summer differs from the state of true hibernation. Therefore, in studies designed to determine the influence of low temperature on arbovirus infection in bats, all experiments were initiated in the fall of the year.

Since it seemed unlikely that all bats within a population naturally infected with arboviruses would be in exactly the same stage of infection at the time of entrance into hibernation, experiments were designed so that the relationship between time of virus inoculation and exposure to a cold environment varied among different groups. Detailed results of studies on the influence of environmental temperature on experimental JBE and SLE virus infection in various bat species appear elsewhere [SULKIN et al., 1966a]. Table VIII presents condensed data from certain experiments with big brown bats (E. fuscus) and little brown bats (M. lucifugus) inoculated with JBE virus and subjected to low temperature at different intervals to represent: (1) animals which might enter hibernation at the peak of the infectious cycle when virus is present in blood and other tissues; (2) animals which might enter hibernation soon after becoming infected and before infection develops to a demonstrable level, and (3) the uninfected bat which might acquire infection during the period of hibernation from an infected mosquito seeking a blood meal.

To determine the influence of low temperature on an established JBE virus infection, big brown bats (E. fuscus) maintained at 24 °C were inoculated subcutaneously with 150 WMICLD$_{50}$ of the OCT-541 strain and held an additional 8 days at room temperature prior to transfer to 8 °C (table VIII, group A). On the day of transfer to the coldroom 23 animals were tested for viremia; 20 had blood titers ranging from 1.0 to 3.5 log units, indicating that the majority of bats transferred into the cold at this time were actively infected with JBE virus. Beginning 3 days after transfer to 8 °C, groups of bats were sampled at intervals to determine the location of virus in the cold-exposed animals. During a period of 70 days at 8 °C, 66 bats were tested for viremia; 26 of these were sacrificed and brown fat, brain and kidney as well as blood were assayed. Infection rates and virus titers in blood and brown fat only are presented since virus was demonstrated in the kidney tissue of 3 and the brain tissue of 1 of the 26 bats sacrificed. Viremia was demonstrable in 31 bats tested during the entire period in the cold. Titers ranged from 1.0 to 3.5 during the first 2 weeks at low temperature, dropping to a range of 1.0–1.3 in bats tested 5 and 8 weeks after transfer to 8 °C. JBE virus was present in the brown fat of 14 of the 26 bats sacrificed out of the cold environment and levels of virus in this tissue remained relatively high throughout the period in the cold. Although it is not evident from the manner in which these data have been condensed for this presentation, virus was demonstrable in the brown fat of 5 animals which were not viremic at the time of sacrifice and in which virus could not be detected in brain or kidney. Four of these animals, in which brown fat alone harbored virus, were among the group sacrificed 60–78 days postinoculation, following 52–70 days at 8 °C. Such data support the hypothesis that interscapular brown adipose tissue could sustain foci of JBE virus infection in the hibernating bat.

The demonstration of JBE virus in bats tested 60–78 days after inocula-

Table VIII. Demonstration of JBE virus (strain OCT-541) in blood and brown fat of *Eptesicus fuscus* and *Myotis lucifugus* inoculated subcutaneously and maintained at various environmental temperatures[1]

Experiment[2]	Environmental temperature, °C	Days after inoculation	Virus demonstrated in	
			blood	brown fat
A	24	8	20/23[3] (1.0–3.5)[4]	
		11–14	10/16 (1.0–3.5)	2/4 (1.2)
	24–8[5]	17–21	11/22 (1.0–3.5)	4/8 (1.0–3.5)
		29–42	5/17 (1.0)	2/6 (3.1–3.5)
		60–78	5/11 (1.0–1.3)	6/8 (1.0–3.5)
		48–62 [34][7]	4/7 (1.0–1.5)	
	24–8–24[6]	62–73 [51]	4/9 (1.0–2.1)	
		89–104 [80]	8/31 (1.0–3.5)	
B[8]		3–6	12/16 (1.0–1.7)	5/16 (1.0)
	5	9–14	3/14 (1.0)	1/14 (1.0)
		17–21	1/12 (1.0)	0/12
		24	3/6 (2.5–2.7)	3/6 (1.2–2.5)
	5–24[9]	27	6/8 (1.0–2.0)	5/8 (1.0–2.2)
		31	0/3	0/3
C[10]		8	0/6	
	8	16	0/10	0/4
		22	0/10	0/4
		31	0/10	0/4
		95	14/14 (1.0–4.0)	
	8–24[11]	99	10/11 (1.5–4.0)	
		108	2/2 (2.0–2.3)	

1 Modified from S<small>ULKIN</small> *et al.* [1966a]. Virus inoculum approximately 150 $WMICLD_{50}$.
2 For a description of the experimental groups, see pp. 41–44.
3 Number infected per number tested.
4 Negative log $WICLD_{50}$.
5 Bats held at 24 °C for 8 days prior to transfer to 8 °C.
6 Bats held at 24 °C for 8 days, followed by varying periods at 8 °C before transfer back to 24 °C.
7 Figure in brackets indicates number of days held at 8 °C.
8 *M. lucifugus* used in this experiment.
9 Bats held at 5 °C for 21 days before transfer to 24 °C.
10 Bats held at 8 °C for 72 h prior to virus inoculation.
11 Bats held at 8 °C for 92 days before transfer to 24 °C.

tion (52–70 days after transfer to 8 °C) indicates that, in many cases, cold exposure prolongs the period during which virus can be recovered from bats following subcutaneous inoculation. When experimentally infected *Eptesicus* are maintained at 24 °C, JBE virus is readily recovered from animals tested 2–14 days postinoculation [SULKIN et al., 1963], but subsequently the incidence of virus isolations becomes erratic, presumably due, at least in part, to the development of specific antibodies [SULKIN et al., 1966b]. The recovery of JBE virus from a significant number of bats over a period of 78 days following inoculation, in concentrations comparable to those obtained at the peak of the infectious cycle in the active bat [SULKIN et al., 1963], suggests that the mechanism responsible for limiting JBE infection in animals at 24 °C is suppressed by exposure of the animals to low temperature.

After varying periods at 8 °C, groups of bats which had been placed in the cold when actively infected with JBE virus were transferred back to room temperature and tested at intervals for presence of virus in the blood. Seven bats were transferred to 24 °C after 34 days in the cold, and over a period of 6–20 days virus was demonstrated in concentrations ranging from 1.0 to 1.5 log units in the blood of 4 animals. Similarly, following a period of 51 days at 8 °C, virus was demonstrated in the blood of 4 of 9 bats tested 3–14 days after transfer to room temperature. The remaining bats were held for a total of 80 days at 8 °C and, upon transfer back to 24 °C, 8 of 31 animals tested 1–16 days later were found to be viremic. In general, the bats in which JBE virus was demonstrated following transfer from 8 to 24 °C were those bled within 1 week of removal from the cold. When only these animals are considered, there is little difference in the ability to demonstrate persistent or recurrent infection in animals held in the cold for as long as 80 days. Failure to recover virus from animals tested more than 7 days after arousal from hibernation may be due to a rapid production of specific antibodies. Correlated studies concerned with antibody response in bats infected with JBE virus and held for extended periods at low temperature have shown that significant levels of neutralizing antibodies develop within 7–14 days after transfer to 24 °C [SULKIN et al., 1966b].

When little brown bats maintained at room temperature were inoculated with 150 $WMICLD_{50}$ of the Nakayama strain of JBE virus and transferred immediately to 5 °C (table VIII, group B), there was evidence of virus multiplication, particularly within the first 6 days of observation. 12 of 16 bats sacrificed during this period had blood virus titers of 1.0–1.7 log units; virus was present in the brown fat tissue of 4 of these viremic bats and a trace amount of virus was demonstrable in interscapular brown fat of another bat which was not viremic at the time of harvest. The infections appeared to be suppressed subsequently, with only trace amounts of virus demonstrable in blood tested after the 6th day. Experimental JBE infection, suppressed by maintenance of the infected *Myotis* at 5 °C, was rapidly activated by transfer of the bats to room temperature. Within 3 days virus concentrations in blood and brown fat of some bats had reached 2.5–2.7 log units, indicating significantly more intense infections than had been demonstrated in any of the infected bats sacrificed from the cold. Infection was demonstrated in 6 of 8 bats sacrificed 6 days after transfer to room temperature, whereas no virus could be demonstrated in blood or brown fat from the 3 remaining bats sacrificed on the 10th day (31 days after inoculation).

In another series of experiments (table VIII, group C) bats were acclimated to low temperature (8 °C) for 72 h before receiving virus to determine if those in deep hibernation in nature could become infected by the bites of mosquitoes present at the site of hibernation and could sustain the infection throughout the period of winter sleep. Big brown bats netted in September were held at 8 °C for 72 h prior to subcutaneous inoculation of 150 WMICLD$_{50}$ of the OCT-541 strain of JBE virus. Torpid bats were removed from the coldroom in small groups, inoculated and returned immediately to the cold environment. Beginning 8 days after inoculation, groups of animals were tested for viremia at weekly intervals for 31 days. In addition, other groups were sacrificed periodically, and brown fat, brain and kidney, as well as blood, were assayed for JBE virus. It was not possible to demonstrate JBE virus in blood or other tissues of these animals during the 31-day period following virus inoculation. These results differ significantly from those obtained in previous experiments in which *Myotis,* inoculated with JBE virus while in an active state at 24 °C, supported multiplication of JBE virus for periods after transfer to the cold. It would appear that either the 72-hour period of acclimatization to 8 °C prior to inoculation prevented replication and subsequent recovery of virus from animals in the cold, or that the response of this bat species to low temperature is so rapid that even had active bats been inoculated and then placed at 8 °C, the slow-down of the biological activities responsible for virus replication would have occurred so quickly that JBE virus could not have been demonstrated in tissues from these animals. LAMOTTE [1958] reported failure to recover JBE virus from the blood of *Eptesicus* in an experiment carried out under the latter conditions. Surviving bats were held at 8 °C for a total of 92 days before transfer to room temperature. Although no virus had been demonstrated in bats from this group during the first 31 days at 8 °C, it can be seen that 2–16 days after transfer to 24 °C significant numbers of animals were circulating virus in concentrations ranging from 1.0 to 4.0 log units.

The unusual thermoregulatory mechanisms of certain bat species offers the unique opportunity of studying *in vivo* the characteristics of a particular virus infection in these animals over a wide range of body temperatures. The only limit to such experiments are those temperatures above or below the range in which the animals are able to survive. The previous studies on JBE virus infection in big brown bats maintained at 24 °C have shown that at this temperature the virus reached titratable levels in blood within 48 h and in brown fat tissue within 72 h. Subsequently, virus could be demonstrated in the brain and kidney tissue of certain bats although none of the animals showed symptoms of encephalitis; hematoxylin-eosin-stained sections of infected brown fat, brain and kidney tissue exhibited no pathological alterations attributable to viral necrosis. Since the low temperature studies described above had shown that under certain circumstances JBE virus infection is dramatically suppressed in the cold-exposed, inactive bat, it seemed of interest to determine if, under similar circumstances, infection would be intensified by the increased body temperature and metabolic rate induced by maintaining the animals at 37 °C. The big brown bat *(E. fuscus)* was used for these experiments because of the observed stamina of this species under various laboratory conditions. Bats were placed at 37 °C 3 days prior to the initiation of experiments to determine the mortality rate of this species under these conditions and to provide a period of acclimatization to the higher temperature. Although these experiments were initiated in October, the bats survived very well at 37 °C. None of the animals died during the initial 3-day period and, subsequently, the mortality rate was no greater than that observed in experiments with this species at 24 °C. Bats were inoculated subcutaneously with approximately 150 WMICLD$_{50}$ of the OCT-541 strain of JBE virus and returned immediately to 37 °C. At subsequent intervals blood, brown adipose tissue, brain and kidney were assayed for virus by the intracerebral inoculation of weanling mice. Assay of blood samples obtained between 24 and 96 h after virus inoculation indicates that the incidence of viremia and the concentrations of virus in the blood rose sharply and reached maximum levels during this period; a dramatic suppression of the viremic state occurred between the 4th and 7th days. Only trace amounts of virus were detected in the brown fat, brain and kidney tissue despite the presence of significant levels of virus in corresponding bloods at the time of harvest. During a period of 7–22 days after virus inoculation only trace amounts of virus were present in the blood and the incidence of tissue involvement remained low. No signs of encephalitis were observed in bats inoculated subcutaneously and maintained at 37 °C.

To determine if virus would multiply in the tissue of bats at 37 °C if introduced directly into a particular tissue, bats acclimated to 37 °C for 72 h were inoculated intracerebrally with 150 $WMICLD_{50}$ of the OCT-541 strain of JBE virus and were maintained at 37 °C. The course of viremia in these animals was similar to that observed in bats inoculated subcutaneously and maintained at 37 °C – virus involvement of brown adipose tissue and kidney was quite low. Inoculation of virus directly into the brain did increase the incidence of infection in this tissue. Virus was demonstrated in appreciable titer in the brains of bats sacrificed 7 days postinoculation, yet none of these animals showed signs of encephalitis and sections of brain tissue revealed little or no damage due to viral invasion and multiplication. In general, the course of JBE virus infection in bats maintained at 37 °C was not as intense or widespread as in the intracerebrally-inoculated bats maintained at 24 °C.

Influence of Serial Passage in Bats Maintained at Low and High Temperatures on the Virulence of JBE Virus

During certain seasons it is possible that mosquito-bat-mosquito cycles of transmission could occur in an environment in which virus replication in both vector and host takes place at a relatively low temperature [LAMOTTE, 1958; CHAMBERLAIN and SUDIA, 1961]; during summer months cycles of transmission could occur in populations of bats roosting in environments with temperatures of 37 °C or higher [GRIFFIN, 1958; HENSHAW, 1970]. The development of an attenuated strain of JBE virus by serial passage in hamster kidney cell (HKC) cultures at 24 °C [ROHITAYODHIN and HAMMON, 1962] suggested that strains of this virus circulating in bat populations in low-temperature environments might be less pathogenic for man than those that undergo alternate cycles in mosquitoes and homoiothermic hosts. Serologic evidence of JBE virus infection in certain areas of northern Hokkaido, Japan, suggested the circulation of virus strains of altered virulence, possibly due to climatologic factors [MIURA and KITAOKA, 1955; SHIRAKI, 1966]. Also, in studies of the relation of mosquito vectors to winter survival of encephalitis viruses in the United States, strains of WEE virus isolated from naturally infected mosquitoes captured during January through March had characteristics of attenuation in that they were nonpathogenic for mice and poorly immunogenic for chickens [REEVES et al., 1958a]. Therefore, studies were undertaken to determine if serial passage of JBE virus in bats maintained at 24 and 37 °C would produce virus populations with characteristics different from the original virus strain [MIDDLEBROOKS et al., 1969]. A strain of JBE virus (OCT-541) was passed serially in big brown bats maintained at 24 or 37 °C, and the virus populations obtained were characterized and compared with those developed by serial passage in HKC cultures incubated at the same temperatures. Virus populations developed by passage in HKC cultures at 24 °C showed significant loss in virulence for mice as reported previously [ROHITAYODHIN and HAMMON, 1962]. Passage of this virus strain in bats maintained at 24 °C, however, did not result in any significant reduction in mouse pathogenicity. These results indicate that circulation of arboviruses in bat populations in low-temperature environments would not produce avirulent strains. Serial passage of JBE virus at 37 °C in either HKC cultures or bats did not affect virulence but did influence the ability of the virus to multiply at these temperatures. Passage in HKC cultures at 37 °C slightly increased the growth of the virus at 37 and 40 °C. Conversely, passage in bats kept at 37 °C to maintain body temperatures at 37 °C or higher seemed to de-adapt the virus for growth in HKC cultures at 40 °C. These data indicate that the effect of temperature on certain characteristics of this strain of JBE virus varies depending on whether serial passage is carried out in HKC cultures or in bats and suggest that studies on the influence of temperature on viruses replicating in bats whose body temperatures can be regulated experimentally might contribute to a better understanding of the mechanism of temperature-induced virus variation.

The Immune Response of Bats to Experimental JBE Virus Infection

Extensive studies in Japan on the detection of immune response following overt and inapparent JBE virus infections in man and experimental and natural infections in birds and swine had demonstrated that the patterns of complement fixing, hemagglutination-inhibiting and neutralizing antibody responses varied among these different species [BUESCHER et al., 1959a, b; SCHERER et al., 1959a, b]. Since ecological and epidemiological studies rely heavily on serological surveys in tracing the occurrence of virus infections in human and animal populations, investigation of the nature of the immune response of bats to JBE virus infection was needed to determine the serological test which would be most sensitive in detecting evidence of infection with this agent in bat populations in nature. Also, the persistence and

inapparency of experimental JBE virus infection in bats suggested the presence in these animals of a mechanism which prevented virus replication from
injuring host cells but failed to completely eradicate viable virus; thus, the
effect of specific antibody on the course of experimental JBE virus infection
in bats needed to be determined. In addition, antibody studies with this host
provided an opportunity to obtain information concerning the influence of
temperature on immunoglobulin-forming mechanisms.

Big brown bats *(E. fuscus)* were used because much information had
already been obtained concerning the course of experimental JBE virus
infection in these animals [SULKIN *et al.,* 1963, 1966a]. Also, this bat species
adapts well to life in captivity and can withstand repeated cardiac bleedings
at 1- to 2-week intervals. Results obtained in these initial studies on the
immune response of *Eptesicus* to experimental infection with the OCT-541
strain of JBE virus [SULKIN *et al.,* 1966b] have been summarized for the
present discussion. Complement-fixation (CF), hemagglutination-inhibition
(HI), and neutralization (N) tests were done according to methods used in
serological analyses of serum samples from humans, birds and swine in
immunologic studies of JBE virus in Japan [BUESCHER *et al.,* 1959a]. In
addition, an *in vitro* microassay plaque-reduction technique using chick
embryo cells grown in leucite hemagglutination plates [MIURA and SCHERER,
1962] was also evaluated for the detection of neutralizing antibodies.

In the initial experiments plasma samples obtained from bats over a
period of 1–10 weeks following a single, infective dose (150 WMICLD$_{50}$)
of JBE virus (OCT-541) were assayed for CF and HI antibodies using
antigens prepared from suckling mouse brains infected with the same virus
strain. No clear-cut evidence of fixation of complement or of inhibition of
hemagglutination was observed with any of the samples tested. The bats
from which these plasma samples were obtained had been tested for viremia
2–5 days postinoculation and thus were known to have been infected; also,
a significant number of these animals did develop neutralizing antibodies
measurable by the weanling mouse intracerebral N test. That the experimentally infected bat failed to produce CF antibodies was not unusual since
this technique had proved to be unsuitable for the detection of JBE virus
antibodies in birds and inferior to HI and N tests for serological analyses
of swine sera in Japan [BUESCHER *et al.,* 1959b; SCHERER *et al.,* 1959b].
Failure of experimentally infected bats to produce antibodies against JBE
virus demonstrable by the HI technique, however, was unexpected. When
repeated attempts to demonstrate HI antibodies in plasma from bats infected by a single dose of JBE virus failed, pooled plasma samples from bats

given multiple doses of virus over a period of several months were included in the HI tests. These hyperimmune plasma pools had N indices of 100–1,000, yet none specifically inhibited hemagglutination by JBE virus antigen.

Throughout the HI studies with bat plasmas an occasional sample would inhibit hemagglutination at low dilution (1:10 to 1:20). However, these low HI titers did not correlate with the presence of neutralizing antibodies in plasma samples and thus could not be interpreted as evidence of an immune response to JBE virus. In addition, nonspecific hemagglutination by normal bat plasma in low dilution was observed from time to time. In order that plasmas from normal and JBE virus-infected bats might be tested for CF and HI antibodies using reagents other than those prepared in our laboratory, samples were sent to the Yale Arbovirus Research Unit where, through the courtesy of Dr. JORDI CASALS, they were assayed. The results were similar to those obtained in our laboratory: 2 of 10 bat plasmas, all of which had N indices of 100–1,000, reacted in low dilution in the HI test and all failed to exhibit evidence of CF.

Since we were able to demonstrate CF and HI antibodies in rabbits and guinea pigs immunized with the OCT-541 strain of JBE virus, the failure to do so in bats was probably not due to the virus strain but rather reflected a host characteristic. However, in studies with Sagiyama virus, a group A arbovirus, it had been noted that whereas one particular strain induced only a heat-labile N antibody in experimentally infected swine, other strains of this virus stimulated the production of CF, HI and N antibody and all 3 types of antibody could be demonstrated in naturally infected swine [SCHERER et al., 1962]. Therefore, additional attempts were made to demonstrate CF and HI antibodies in *Eptesicus* infected with another strain of JBE virus (JaGAr-01), also a mosquito isolate [OYA et al., 1961]. In this study, bats were maintained at 24 and 37 °C and groups inoculated intracerebrally were included in addition to animals receiving virus via the subcutaneous route. Again, no clear-cut evidence of CF or HI antibody production was obtained, although the N antibody response was equal to that observed in studies with the OCT-541 strain of JBE virus [LEONARD et al., 1968].

Serological evidence of experimental JBE virus infection in bats was most readily demonstrated by the *in vivo* N test using constant-serum versus varying-virus dilutions; LD_{50} and log N indices (LNI) were calculated by the methods of Reed and Muench or Karber [LENNETTE, 1969]. An LNI of 1.7 or greater was considered positive, 1.0–1.7 equivocal, and less than 1.0 negative. Also, the *in vitro* microassay technique [MIURA and SCHERER, 1962] was used in some instances since, generally, plasmas which were positive or equivocal in the *in vivo* test could be detected in dilutions ranging from 1:3 to 1:12 by the *in vitro* method.

In bats maintained at 24 °C viremia became demonstrable in most animals 2–3 days following inoculation and usually persisted for 10–15 days. Neutralizing antibodies developed within about 3–7 weeks. Generally, only about 75% of the bats known to be infected following a single dose of virus developed positive LNI, which usually ranged between 2 and 3 log units; however, virus also disappeared from the blood of bats which never developed a demonstrable immune response. In a few instances, virus and antibody were demonstrated in blood simultaneously. Such observations suggest a virus-antibody complex which dissociates readily or antibody which combines with virus but fails to neutralize infectivity. In this regard, it is of interest that the constant-virus versus varying-serum dilution method could not be used in the *in vivo* N test since dilution in excess of 1:2 significantly reduced the neutralizing capacity of bat antibody against JBE virus.

In studies comparing the course of JBE virus infection in bats maintained at 37 °C with that observed in room-temperature bats, it had been found that although viremia occurred more rapidly and reached higher levels in bats held at 37 °C, the duration of the viremic phase was shorter and there was little evidence of virus multiplication in brown fat, brain or kidney [SULKIN *et al.*, 1966a]. Bats had high levels of virus in blood 2–4 days postinoculation and by the 7th day all animals were negative or had only trace amounts of virus in blood. Since this limitation of infection might be due to a rapid response of antibody-forming mechanisms in bats maintained at a high temperature, neutralizing antibody levels were determined in JBE virus-infected bats held at 37 °C. These animals developed equivocal LNI's within 2 weeks which rose to positive levels by the third week. Despite the increased rate of antibody formation in most of the bats maintained at 37 °C, about 25% of the animals never developed demonstrable levels of neutralizing antibody, the same as observed in experiments at 24 °C. Also, the N indices of 37 °C bat plasmas were no higher than those of plasmas from bats kept at 24 °C.

Studies on the influence of low temperature on the course of experimental JBE virus infection in bats had shown that when infected animals were transferred to the cold (8 °C) at the peak of the infection, active virus persisted in various tissues of the dormant animals for long periods [SULKIN *et al.*, 1966a]. Thus, whereas JBE infection in bats maintained at 24 °C is apparently suppressed by antibody formation in approximately 3 weeks and in even less time in animals held at 37 °C, virus can be isolated from the cold-exposed infected animal for at least 8 weeks. These observations suggested a suppression of the immune mechanisms of the bat at 8 °C despite prolonged exposure to the antigen (JBE virus) present in various tissues. In order to determine if bats experimentally infected with JBE virus would form specific N antibody during prolonged periods in simulated hibernation, bats were inoculated subcutaneously and maintained at 24 °C for 8 days to allow infection to become established at which time the animals were transferred to a cold environment (8 °C). During the weeks in which the bats were held at 8 °C occasional animals were checked for viremia and returned to the cold, and others were sacrificed so that tissues other than blood could be assayed for virus and the immune status of the animal determined. After almost 9 weeks at 8 °C several bats were transferred back to room temperature and bled at intervals for evidence of viremia and N antibody. Virus was demonstrated in one or more tissues of bats sacrificed 3–8 weeks after transfer to 8 °C and the patterns of the infection and levels of virus in the various tissues were similar to those observed in previous low temperature studies [SULKIN et al., 1966a]. There was no evidence of anti-JBE virus N antibody in any of the plasma samples obtained from cold-exposed bats; no immune response was detectable in at least 50 bats infected with JBE virus and subjected to prolonged periods in the cold during fall and winter months.

Bats were transferred from 8 to 24 °C approximately 9 weeks after virus inoculation and assays for virus and antibody in serial blood specimens obtained over a subsequent period of 16 weeks indicated that the general pattern of infection in these animals consisted of a transient viremia followed by the rapid formation of significant levels of antibody which persisted for at least 11 weeks. The simultaneous presence of JBE virus and specific antibody in a blood specimen obtained 47 days after the animal was transferred out of the cold (more than 15 weeks after inoculation) suggested a recurrent viremia. As procedures for the care and maintenance of bats in the laboratory improved and became increasingly successful in sustaining these animals in the various artificial environments created for experimental purposes, bats survived for longer periods of time. A number of bats survived for 2–3 years despite primary and challenge doses of JBE virus, transfer from 24 to 8 to 24 °C, and numerous cardiac punctures to obtain blood for virus and antibody determinations. These long-term survivors provided opportunities to obtain information concerning the influence of hibernation on the persistence of antibody in bats, the effect of the physiological functions associated with arousal from hibernation on the activation of latent JBE virus infection in these animals, and to demonstrate the occurrence of spontaneous recurrent viremias and susceptibility to reinfection with JBE virus [SULKIN et al., 1966b]. Results obtained in these studies indicated that the immune response of bats to certain arbovirus infections would not completely repress their effectiveness as reservoir hosts. Specific N antibodies, when formed, were not always long-lasting or protective, since evidence of recurrent viremias and susceptibility to reinfection was obtained and the simultaneous demonstration of virus and antibody in bat blood was not uncommon. These observations are particularly significant in view of the recorded information on the longevity of this animal species. Recoveries of banded bats after more than 20 years have been reported [GRIFFIN, 1958; GRIFFIN and HITCHCOCK, 1965]; since these animals do not succumb to experimentally induced arbovirus infection, it seems likely that an infected bat could maintain the agent in nature throughout its lifetime.

Natural JBE Virus Infection in Bats

The data obtained in the experimental studies discussed in the previous sections pointed up the potential effectiveness of bats as reservoir hosts for JBE virus and when considered along with certain characteristics of the physiology and behavior of these animals, such as the phenomenon of hibernation common to many species and the seasonal migratory movements of others, suggested that natural JBE virus infection in bats could contribute to overwintering of this agent in certain locales or to reintroduction of the virus into endemic areas each spring. Although certain phases of the ecology of JBE virus in Japan had been well defined, the means by which this agent persists in nature throughout the winter and gives rise to recurrent epidemics had only been hypothesized [SCHERER and BUESCHER, 1959]. Thus, field studies to investigate the role of bats in the ecology of JBE virus in Japan were initiated in 1963 [SULKIN et al., 1970; MIURA et al., 1970].

Certain areas within 3 main regions of Japan (Kyushu, Honshu and Hokkaido) were chosen as survey sites because of the feasibility of collecting bats in these locales (fig. 1). The bat population in Hokkaido was sampled at 7 different locations throughout the island, whereas 16 collection sites

Fig. 1. Map of Japan showing distribution of cases of Japanese B encephalitis during 1963–1965 together with locations of bat collections. The encircled numbers indicate sites from which bats infected with JBE virus were obtained.

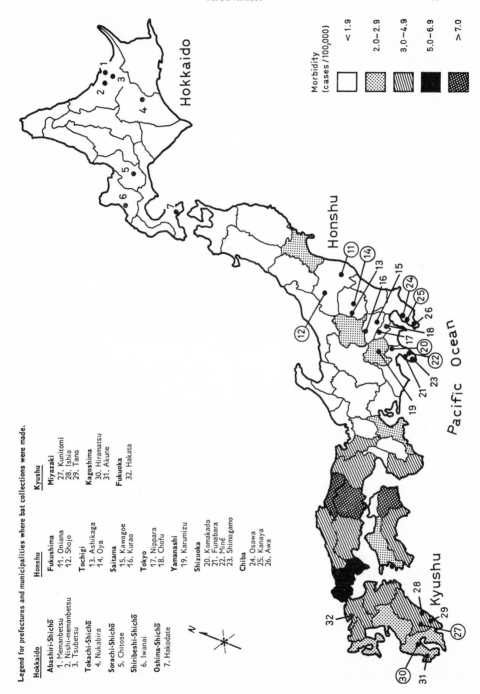

Legend for prefectures and municipalities where bat collections were made.

Hokkaido	Honshu	Kyushu
Abashiri-Shichō	**Fukushima**	**Miyazaki**
1. Memanbetsu	11. Oniana	27. Kunitomi
2. Nishi-memanbetsu	12. Shojo	28. Ishio
3. Tsubetsu		29. Tano
	Tochigi	
Tokachi-Shichō	13. Ashikaga	**Kagoshima**
4. Nukabira	14. Oya	30. Hiramatsu
		31. Akune
Sorachi-Shichō	**Saitama**	
5. Chitose	15. Kawagoe	**Fukuoka**
	16. Kurao	32. Hakata
Shiribeshi-Shichō		
6. Iwanai	**Tokyo**	
	17. Nippara	
Oshima-Shichō	18. Chofu	
7. Hakodate		
	Yamanashi	
	19. Karumizu	
	Shizuoka	
	20. Komakado	
	21. Funabara	
	22. Miné	
	23. Shimogamo	
	Chiba	
	24. Osawa	
	25. Kanaya	
	26. Awa	

Morbidity
(cases /100,000)

< 1.9
2.0 – 2.9
3.0 – 4.9
5.0 – 6.9
> 7.0

were established in Honshu in an area concentrated around Tokyo. A quarry
at Miné, Shizuoka Prefecture, was visited on 19 occasions from August 1963
to October 1965. In Kyushu, bats were obtained from 6 different collection
sites, two of which (Kunitomi, Miyazaki Prefecture, and Hiramatsu, Kago-
shima Prefecture) were visited on more than one occasion. At Kumitomi, for
example, 7 different collections were made at intervals of 1–6 months from
October 1963 to July 1965. Figure 1 also shows the distribution of cases of
JBE during the period when the bat collections were made – retrospective
information obtained after all bats had been collected. Unfortunately, no
collections were made in the southern part of Honshu where most cases of
encephalitis occurred. However, since the flight range of certain bat species
may be 50–60 miles during night-time forages for food, and several hundred
miles during seasonal migrations, the sites from which bats were netted need
not bear a close relation to the epidemic areas.

Collections were spread over a period from July 1963 to December 1965,
so that significant numbers of animals might be obtained at intervals
throughout the year to determine if virus was circulating in bats only during
the months when other natural hosts and mosquito vectors could be found
harboring the agent or if infection could be demonstrated in bats during
months when extensive surveys by other investigators indicated the virus
was hidden in nature [SCHERER and BUESCHER, 1959]. The total number of
bats obtained in each of the 3 regions of Japan is shown in table IX according
to species and habitat. With the exception of a few collections made in dwell-
ings, the remainder of the bats were obtained from caves, quarries, mines
and tunnels. In Honshu, the 988 bats collected from August 1963 to December 1965 comprised 8 species. In Kyushu, the southern region of Japan,
648 bats of 6 species were collected from October 1963 through November
1965 and, in Hokkaido, 298 bats including 5 different species were collected
from July 1963 through September 1965. The best seasonal distribution was
obtained in Honshu where the total number of bats netted was almost
equally distributed over the 4 seasons. Although the number of bats collected
during the winter in Kyushu was small, adequate samplings of the bat
population were obtained during the other 3 seasons of the year. Among the
bats collected in Honshu and Kyushu during the months of April, May and
June were 112 gravid females. In Hokkaido, where fewer bats were netted,
the seasonal distribution of collections was more uneven than in either Honshu or Kyushu; over half the total number of bats obtained were netted
during the summer months, whereas no animals were captured during the
winter seasons.

Table IX. Bat collections in Japan (1963–1965)[1]

Species	Habitat	Number obtained in each collection area			Total
		Honshu	Kyushu	Hokkaido	
Miniopterus schreibersii	lava cave, quarry, shelter, water tunnel	692	475		1167
Rhinolophus cornutus	cave, quarry, mine, sea cave	195	72	70	337
Vespertillio superans	dwelling	20		90	110
Myotis macrodactylus	water tunnel, quarry, mine, cave, shelter, sea cave	2	40	67	109
Rhinolophus ferrum	quarry, lava cave, shelter, mine, cave	34	25	29	88
Myotis mystacinus	dwelling			42	42
Pipistrellus abramus	dwelling	21	20		41
Plecotus auritus	cave, lava cave	22			22
Myotis nattereri	water tunnel, mine		16		16
Murina leucogaster	lava cave, quarry	2			2
Total		988	648	298	1934

1 From SULKIN and ALLEN [1970].

For details concerning collection, processing and assay of bat tissues and identification of virus isolates, the reader is referred to previous publications [MIURA *et al.,* 1970; SULKIN *et al.,* 1970]. Specimens obtained included whole blood, interscapular brown fat, brain, kidney, spleen and salivary glands for virus assay and plasma for antibody studies. All specimens were quick-frozen and stored at –60 °C awaiting shipment to Dallas or assay in the laboratories in Tokyo. Emphasis was placed on the assay of blood specimens from the bats collected in Japan since in the experimentally infected bat JBE virus was demonstrable in the blood for prolonged periods [SULKIN *et al.,* 1963, 1966a] and because viremia is an essential characteristic of infection in animals suspected of being natural hosts for the arboviruses. In addition, because of our interest in brown fat as a focus of infection in the persistently infected bat, pooled suspensions of this tissue from selected groups of bats were assayed as were all tissues from the bats from which JBE virus was isolated from the blood.

Table X. Seasonal distribution of collections of the two bat species from which strains of Japanese B encephalitis (JBE) virus were isolated[1]

Season collected	Number of specimens and isolations from				
	Miniopterus schreibersii		*Rhinolophus cornutus*		Totals[2]
	Number tested	Number of isolations	Number tested	Number of isolations	
Spring	400	1	19	3	4/419
Summer	266	2	68	2	4/334
Fall	339	12	101	0	12/440
Winter	134	1	79	3	4/213
Totals	1,139	16	267	8	24/1,406

1 From SULKIN *et al.* [1970].
2 Number of isolations of JBE virus per number of specimens of bat blood tested.

The encircled numbers in figure 1 indicate sites from which bats infected with JBE virus were obtained. No virus was recovered from bats netted at 7 different locations throughout Hokkaido, the northernmost part of Japan. 19 virus isolates, 14 of which have been identified as JBE virus, were obtained from bats collected in Honshu. Two isolates show some reactivity with group B arboviruses in the HI test but have not been specifically identified as JBE virus. Three isolates remain to be characterized. Additional virus isolates included a strain of JBE virus isolated from the pooled brown fat tissue of bats *(Miniopterus schreibersii)* netted in Miné, Shizuoka, in October 1964, and an uncharacterized agent isolated from the salivary glands of a bat collected in November 1964 from which JBE virus had been isolated from the blood. 14 virus isolates were obtained from bats netted in Kyushu, and 10 of these were identified as strains of JBE virus. Two isolates are apparently group B-related agents, and two other isolates are as yet uncharacterized.

In table X, the 24 strains of JBE virus recovered from the blood of bats collected in Honshu and Kyushu during 1963–1965 are tabulated according to species of bat from which they were recovered and the season of the year in which the bats were collected. 16 of the strains were isolated in assays of blood specimens from 1,139 *M. schreibersii* and 8 strains were recovered from the 267 *Rhinolophus cornutus* tested for viremia. JBE virus was isolated from bats during each of the 4 seasons of the year. The highest incidence of natural JBE virus infection was demonstrated in groups of *Miniopterus* netted in the fall; 12 isolates were obtained from the 339 specimens of blood tested.

Prompted by studies on experimental JBE virus infection in bats [LA-MOTTE, 1958; SULKIN *et al.,* 1963, 1964, 1966a, b], a series of annual surveys initiated in 1967 to provide a continual surveillance for the presence of JBE virus and its vectors in the Taipei area of Taiwan was extended to include investigation of mosquitoes and bats found in caves [CROSS *et al.,* 1971]. The collection of cave specimens began in late December 1969, and only 3 bats *(Hipposideros armiger)* were captured during the remainder of that year. However, JBE virus was isolated from the blood of one of these collected on December 28, 1969. Bat collections in 1970 consisted of 3 *H. armiger,* 5 *H. bicolor,* and 66 *M. schreibersii.* No virus isolates were obtained from the *Hipposideros* sp. but JBE virus was isolated from the blood of two *Miniopterus* which had been collected on August 20, 1970, from a cave in Hsinchuang. JBE virus was also isolated from a pool of 9 *Culex annulus,* the major vector of this virus on Taiwan, collected on the same day and from the same cave as the virus-infected bats.

Plasma samples from 1,459 of the bats collected in Japan were tested for antibodies against JBE virus by the microculture plaque-reduction method of MIURA and SCHERER [1962]. Unheated plasma was tested against 100 plaque-forming units of virus and results were expressed as the serum dilution titer (SDT) that reduced plaque counts 80%. Plasma samples diluted 1:4 or more which reduced plaque counts 80% or more were regarded as positive. The results are summarized in table XI [MIURA *et al.,* 1970]. The incidence of neutralizing antibodies against JBE virus in bats collected in Honshu and Kyushu was about equal (8%), whereas only 5 of 198 bats (3%) collected in Hokkaido had significant SDT's. The incidence of neutralizing antibodies in plasma samples from *M. schreibersii* and *R. cornutus,* the two species from which all JBE virus isolations were made, was 5 and 9%, respectively. (Note the demonstration of the simultaneous presence of JBE virus and neutralizing antibody in the blood of a *Miniopterus* netted in Kyushu.) Serologic evidence of JBE virus infection was also demonstrated in several other species, however, suggesting that natural infection of bat populations in Japan with this agent is more widespread than the virus isolation studies indicated. Failure to isolate JBE virus from certain species may well have been due to the small numbers collected. An additional 275 plasma samples from bats netted in Honshu and Kyushu were assayed by the mouse N test. The results obtained with this method correlated well with the *in vitro* data in that 20 (7%) of the samples had positive or equivocal

Table XI. Incidence of neutralizing antibody against JBE virus in bats collected in Japan as measured by the microculture plaque-reduction method[1]

Collection area	Bat species[2]										Totals
	Mi.s	*Rh.c*	*Ve.s*	*My.m*	*Rh.f*	*Pi.a*	*My.my*	*My.n*	*Pl.a*	*Mu.l*	
Honshu	32/529[3]	12/141	10/20	0/2	7/30	1/20			1/22	1/2	64/766 (8)[4]
Kyushu	8[5]/351	8/55		9/39	14/25	0/11		0/14			39/495 (8)
Hokkaido		1/28	0/90	0/31	0/24		4/25				5/198 (3)
Totals	40/880 (5)	21/224 (9)	10/110 (9)	9/72 (12)	21/79 (27)	1/31 (3)	4/25 (16)	0/14	1/22 (5)	1/2	108/1,459 (7)

1 From MIURA *et al.* [1970].
2 Abbreviations of bat species: *Mi.s = Miniopterus schreibersii; Rh.c = Rhinolophus cornutus; Rh.f = Rh ferrum-equinum; My.m = Myotis macrodactylus; My.n = My nattereri; Pi.a = Pipistrellus abramus; Pl.a = Plecotus auritus; Ve.s = Vespertilio superans; My.my = My mystacinus,* and *Mu.l = Murina leucogaster.*
3 Number of specimens of plasma with serum dilution titer (SDT) of 1:4 or greater per number tested.
4 Number in parentheses indicates percent positive.
5 JBE virus was isolated from the blood of one of these bats.

LNI. Levels of neutralizing antibodies in bat plasmas as measured by either method were low; more than 50% of the samples which were positive in the plaque-reduction test had SDT no greater than 1:4, and only 5 of 20 samples which showed serologic evidence of JBE virus infection in the *in vivo* test had LNI's of 1.7 or greater.

To determine if evidence of natural JBE virus infection could be demonstrated in bat plasmas by the HI technique, a method which had failed to detect antibodies in the experimentally infected bat [SULKIN *et al.,* 1966 b], 437 plasmas that had been assayed for neutralizing antibodies by the plaque-reduction method were tested for HI antibodies. Although 41 of these plasmas had neutralizing antibodies, only 2 reacted in low dilution (1:10) with a hemagglutinin made with the Nakayama strain [KASAHARA *et al.,* 1936] of JBE virus. Thus, as with experimentally infected bats, serologic evidence of natural JBE virus infection in bats collected in Japan could not be demonstrated by the HI test. Similar results were reported by WANG *et al.* [1962] who tested serum pools from 228 bats collected in Taiwan for antibodies against JBE virus. Three of 15 pools had low levels of neutralizing antibodies but none were positive in the HI test. Other investigators have reported the demonstration of HI antibodies against group B arboviruses, including JBE virus, in *Pteropus* bats collected in India; however, only one of three positive bloods reacted with JBE hemagglutinin alone and the titer was low, making the specificity of the test questionable [CAREY *et al.,* 1968].

The demonstration of neutralizing antibodies against JBE virus in plasma from bats collected in Japan supported the results obtained in the virus isolation studies [SULKIN *et al.,* 1970] and provided additional evidence that bat populations in Japan are persistently infected with this virus. That the incidence and levels of neutralizing antibodies in bat plasmas were low was not surprising since it had already been shown that not all experimentally infected bats produce demonstrable antibodies against JBE virus, and plasmas from those which do seldom have N indices exceeding 100 [SULKIN *et al.,* 1966 b]. The low incidence of neutralizing antibodies against JBE virus among the large numbers of bats tested indicated that a nonspecific virus-neutralizing substance is not present in bat plasma; the similarity in the characteristics of the immune response of bats naturally infected with JBE virus to that of the experimentally infected bat favors the supposition that specific neutralizing antibodies were measured. Also, other investigators have reported the demonstration of neutralizing antibodies against JBE virus in serum from bats collected in other areas. Evidence of neutralizing antibodies against this agent in sera from *Pteropus* bats in Australia has been reported [ROWAN and O'CONNER, 1957] and, in more recent studies on the ecology of arboviruses in Thailand, neutralizing antibodies against JBE virus were demonstrated in sera from 22 of 245 bats *(Cynopterus brachyotis)* included in the survey [RUSSELL, 1968].

Experimental St. Louis Encephalitis (SLE) Virus Infection in Bats

The initial studies to determine if bats might be involved in the natural history of SLE were concerned with demonstrating the susceptibility of these animals to experimental infection with this agent; 3 strains of SLE virus and 2 bat species *(T. brasiliensis* and *M. lucifugus)* were used [SULKIN *et al.,* 1963]. Strains of SLE virus included a high mouse passage strain (Hubbard) and two low mouse passage strains – a mosquito isolate (M-57-4MB) and one obtained from a naturally infected flicker bird (FB-55-7MB). In all experiments each animal received approximately 150 WMICLD$_{50}$ of virus subcutaneously. Variations in the susceptibility of these bat species to the different strains of SLE virus were observed. The Hubbard and M-57-4MB strains failed to establish demonstrable infection in *Myotis* and were only slightly infective for *Tadarida*. The flicker bird strain, however, induced a low-grade infection in *Myotis,* and in *Tadarida* produced a particularly intense and long-lasting viremia. Also, viral invasion and multiplication was demonstrated in the brown fat and, to a lesser degree, in the brains and kidneys of some animals; yet none of the infected bats showed overt signs of encephalitis.

Since *Myotis* appeared to possess some degree of innate resistance to SLE virus (neither specific antibody nor nonspecific inhibitor substances could be demonstrated in plasma from uninoculated bats), it was concluded that these animals would not be the species of choice in surveys of bat populations for evidence of natural infection with this agent. *Tadarida,* on the other hand, were considered as potentially effective natural hosts for SLE virus.

The low infectivity of certain strains of SLE virus for nongravid Mexican free-tailed bats [SULKIN et al., 1963] together with the availability in nature during May and June of large numbers of gravid *Tadarida* in various stages of the reproductive cycle, provided an opportunity to determine whether the effects of the physiological stress of pregnancy would alter the course of SLE virus infection in this host. Bats in progressive stages of the reproductive cycle, as determined by sacrificing groups of bats from each collection and determining fetal weights [SIMS et al., 1963], were inoculated subcutaneously with the M-57-4MB strain of SLE virus; at subsequent intervals animals were sacrificed and blood, brown fat, brain, kidney and whole fetus were assayed for virus. Low-grade viremia was demonstrable in only about 12% of the bats receiving virus during early, mid or late stages of pregnancy; in a few instances small amounts of virus were detected in brown fat and kidney tissue. A comparison of these results with those obtained in previous studies with nongravid *Tadarida* indicated that the gravid state failed to influence significantly the susceptibility of this species to infection with this strain of SLE virus [SULKIN et al., 1964]. Treatment of *Tadarida* with cortisone also failed to enhance the multiplication of this strain of SLE virus in host tissues.

In experiments with the M-57-4MB strain of SLE virus just described, fetuses from 7 viremic gravid bats were assayed. Virus was detected in the whole fetus obtained from a bat inoculated during the mid-gestation period and sacrificed 8 days later. In subsequent experiments with gravid *Tadarida* infected with the flicker bird strain of SLE virus (FB-55-7MB), which induces a relatively high-titered and long-lasting viremia in this species, no evidence of transplacental transmission was obtained in bats infected during early and mid stages of pregnancy; virus was found to cross the placenta in only 1 of 17 bats infected during the late stage of pregnancy [SULKIN et al., 1964]. In experiments discussed above, transplacental transmission of JBE virus in *Tadarida* was demonstrated in significant numbers of bats, particularly those infected during the late stage of pregnancy, indicating that this mechanism could contribute to the perpetuation of this agent in bat populations. Presumably SLE virus is less likely to be maintained in nature through transplacental passage in bats.

Studies on the influence of temperature on SLE virus infection in bats have not been as extensive as those discussed above for JBE virus. In a single series of experiments, active Mexican free-tailed bats were inoculated with 150 WMICLD$_{50}$ of the high mouse passage Hubbard strain of SLE virus and placed at 10 °C [SULKIN et al., 1966a]. During a subsequent observation period of 20 days virus was demonstrated in 10 of 44 bats sacrificed from the cold; a low-level viremia was demonstrable in 2 bats, whereas in 8 animals virus was detected in brown fat but not in blood, suggesting that virus sequestered in the brown adipose tissue of the cold-exposed bat. Virus was recovered from the blood and/or brown fat of 8 of 21 bats sacrificed 3–6 days following transfer to room temperature after 21 days at 10 °C, indicating an increased rate of viral replication at the higher temperature. Since the Hubbard strain of SLE virus had been shown to be of low infectivity for bats of this species even when maintained in an active state [SULKIN et al., 1963], it was of particular interest that infection with this strain was sustained in *Tadarida* for at least 21 days at 10 °C.

Natural SLE Virus Infection in Bats

Epidemics of SLE in Houston, Texas, in the summer of 1964 followed by outbreaks in Dallas and in Corpus Christi, Texas, in 1966 [LUBY et al., 1969] prompted the initiation of field studies in these areas to determine if bats might be involved in the epidemiology of this disease. Bat populations near Houston and in Corpus Christi were surveyed during 1964–1967 for evidence of natural infection with SLE virus [ALLEN et al., 1970]. Efforts to find colonies of bats in the Houston metropolitan area, where the center of the epidemic was located, were unsuccessful. A large colony of Mexican free-tailed bats was found, however, in Angleton, Brazoria County, about 40 miles south of Houston, a distance within the nocturnal-feeding flight range of this species [DAVIS et al., 1962]. The first collection of bats was made on August

26, 1964, which subsequently proved to be during the week when the epidemic reached a peak with regard to number of cases. In order to determine
if SLE virus could be detected in the bat population subsequent to the
epidemic period and at a time when the agent was no longer demonstrable
in the vector or known reservoir hosts such as birds, we netted additional
groups of bats in September, October, November and December, 1964.
Collections were resumed in April 1965 and continued at about monthly
intervals through January 1966 to determine if SLE virus was circulating
in the bat population in this area during what proved to be a nonepidemic
year. In Corpus Christi, colonies of *T. brasiliensis* were found roosting along
rafters and beams in the ceilings of warehouses in the dock area of the city.
The first collection was obtained on September 11, 1966, and collections
were continued at about 14-day intervals through November, and then
monthly through May 1967. In June 1967 the colonies were disrupted by
officials of the city-county health department. Sufficient numbers of bats
were obtained during each of the 4 seasons of the year in the Houston area
and, although we were not able to collect bats in Corpus Christi after May
1967, we had adequate samples from the spring, fall and winter months – the
seasons in which we were most interested in detecting SLE virus infection
in the bats in this area. The specimens obtained from these animals and the
assay procedures employed to detect virus and antibody were essentially the
same as those established for the field studies in Japan [SULKIN *et al.*, 1970;
MIURA *et al.*, 1970]. Because of the incidence of natural infection with Rio
Bravo virus in populations of *T. brasiliensis* in the southwestern United
States [CONSTANTINE and WOODALL, 1964], procedures for specific identification of viruses were designed to differentiate this agent from SLE virus and
consisted of checkerboard serum N tests with known strains of SLE and Rio
Bravo virus [SULKIN *et al.*, 1966c].

A total of 1,016 Mexican free-tailed bats were collected in Angleton at
intervals between August 26, 1964, and January 28, 1966. Blood specimens
from 659 and spleen tissue from 192 of these animals were assayed in suckling mice; 275 plasma samples were tested for neutralizing antibodies against
SLE virus. Blood specimens from 798 of the 870 bats collected in Corpus
Christi between September 11, 1966, and May 31, 1967, were assayed for
virus, and 388 plasma samples were checked for serologic evidence of natural
SLE virus infection. Table XII lists the isolations of SLE virus from bats
collected in Angleton and Corpus Christi, Texas, during 1964–1967. In
addition to strains of SLE virus isolated from the blood of bats netted during
the height of the Houston epidemic (August 26, 1964), strains of this agent

Table XII. Isolations of St. Louis encephalitis (SLE) virus from *T. brasiliensis* collected in Texas during 1964–1967[1]

Collection area	Date collected	Number of isolations[2]
Angleton	August 26, 1964	2
(Brazoria County)	September 8, 1964	1
	November 11, 1964	1
	April 20, 1965	2[3, 4]
	July 12, 1965	1[3]
	November 23, 1965	4
	December 3, 1965	1
	January 28, 1966	1
Corpus Christi	September 11, 1966	4
(Nueces County)	September 28, 1966	1
	November 9, 1966	1
	November 28, 1966	1
	December 22, 1966	1
	February 22, 1967	1
	April 5, 1967	2
	May 31, 1967	2

1 Modified from ALLEN *et al.* [1970].
2 All isolations from blood except where indicated.
3 Isolations from pooled splenic tissue of 4 bats.
4 Neutralizing antibody against SLE virus demonstrated in pooled plasma samples from bats in these two groups (table XIII).

were recovered from bats obtained in September and November 1964. Records indicate that SLE virus was not detectable in mosquito pools collected after September 20 of that year. Although follow-up studies in Houston showed no evidence of SLE virus infection in mosquitoes, birds, man or sentinel flocks of chicks during the summer of 1965 [PHILLIPS and MELNICK, 1967], additional strains of SLE virus were obtained from the spleen tissue of bats collected in April and July 1965 and from the blood of bats collected in November and December 1965 and in January 1966, indicating the persistence of this virus in populations of Mexican free-tailed bats in this area. Although collection of bats in Corpus Christi was begun late in the epidemic period when the attack rate was declining rapidly, 4 strains of SLE virus were isolated from bats obtained in the first collection on September 11,

1966; this agent was isolated from virtually every subsequent collection from this location. The isolation of SLE virus from bats netted in November and December of 1966 and February, April and May of 1967 demonstrated the persistence of SLE virus infection in bat populations in the Corpus Christi area during this period. Rio Bravo virus was isolated from bats collected in Angleton and in Corpus Christi and several uncharacterized viruses were also recovered from bats netted in the latter location.

The isolation of SLE virus from pooled splenic tissue of bats collected during the spring and summer of 1965 in Angleton, Texas, was of particular interest to us. Unpublished observations in our laboratory with bats experimentally infected with JBE virus had shown that the isolation of virus from the spleens of viremic animals paralleled the detection of virus in the blood and suggested that when blood samples were insufficient to allow testing for both virus and antibody or when the expediency of spleen removal was more desirable than bleeding by cardiac puncture, the presence of virus in the spleen could be used as an index of viremia. Results obtained in our field studies confirmed this observation since the isolation of SLE virus from at least 3 of the 192 bat spleens assayed compared favorably with the isolation of this agent from 10 of 659 blood samples tested. As indicated in tables XII and XIII, neutralizing antibodies against SLE virus were detected in pooled plasma samples of bats in two of the groups from which SLE virus was isolated from pooled splenic tissue. The simultaneous presence of virus and low levels of specific antibody was also observed in bloods from bats experimentally and naturally infected with JBE virus [SULKIN et al., 1966b; SULKIN et al., 1970; MIURA et al., 1970].

The incidence of neutralizing antibodies against SLE virus in plasma samples from bats collected in Texas during 1964–1967 is shown in table XIII. Plasma samples from bats collected in Angleton in December 1964 and in April and October of 1965 were tested in pools of 3–4 plasma samples each; an additional 84 samples from bats collected in December 1965 and January 1966 were assayed individually. An LNI of 1.0 or greater was considered positive, and a positive pool was counted as containing only one positive plasma. By these criteria, at least 25 (9%) of the 275 plasma samples from bats collected in Angleton, Texas, contained neutralizing antibodies against SLE virus. A total of 388 plasma samples from bats collected in Corpus Christi, Texas, were tested and 108 samples (28%) were positive against SLE virus. The incidence of neutralizing antibodies in bats collected in Corpus Christi seemed significantly higher than in those collected in Angleton, possibly due to the fact that the samples from bats collected in Corpus

Table XIII. Incidence of neutralizing antibodies against SLE virus in *T. brasiliensis* collected in Texas during 1964–1967[1]

Collection area	Date collected	Number of plasma samples	Positive[2]	
			Number	%
Angleton	December 1964	35[3]	0	
(Brazoria County)	April 1965	88[3]	7[4]	8
	October 1965	68[3]	7	10
	December 1965	40	4	10
	January 1966	44	7	16
Total		275	25	9
Corpus Christi	September 1966	125	48	38
(Nueces County)	October 1966	131	40	31
	November 1966	24	1	4
	February 1967	24	0	
	April 1967	84	19	23
Total		388	108	28

1 From ALLEN *et al.* [1970].
2 Minimum number of plasma samples with log neutralization indices (LNI) of 1.0 or greater. A positive pool was counted as containing only one positive plasma.
3 Tested in pools of 3–4 plasma samples each.
4 SLE virus was isolated from pooled splenic tissue of bats included in two of these positive pools (table XII).

Christi were tested individually whereas the greater proportion of the samples from bats obtained in Angleton were pooled and a positive pool of 3–4 samples was counted as containing only one positive plasma.

Selected groups of plasma samples from bats netted in Angleton and Corpus Christi were tested for the presence of HI antibodies. Hemagglutinins made from strains of SLE virus isolated from *Tadarida* were used in the tests in addition to hemagglutinins made from strains of mosquito, bird and human origin. Although these hemagglutinins reacted in high titer with mouse, guinea pig and rabbit antiserum against SLE virus, bat plasmas known to contain neutralizing antibodies against this agent did not inhibit hemagglutination by these antigens. The detection of antibodies against an SLE virus hemagglutinin in bat serum samples has been reported by other investigators, however; two serum pools from a total of 37 bats (species unidentified) collected in New York state inhibited hemagglutination by SLE virus antigen in titers of 1:160 and 1:320, respectively [WHITNEY, 1963]. The author suggested that these high HI titers with SLE antigen could have been due to the presence of antibodies against Rio Bravo virus in the bat sera.

In a survey in Wisconsin in 1960, neutralizing antibodies against SLE virus were detected in serum samples from 12 of 127 *Eptesicus,* but SLE virus was not isolated from blood, heart, liver or lung tissues from bats collected at the same time [RUEGER *et al.,* 1966].

Other Group B Virus Infections in Bats

'Bat' viruses. Seven viral agents, isolated from the salivary glands (and in certain instances from other tissues also) of naturally infected bats, have been classified provisionally as group B arboviruses on the basis of antigenic relatedness alone, since there is no firm evidence that any of these agents are transmitted by arthropods. Rio Bravo virus, the first of the 'bat' viruses to be identified, was isolated from the salivary glands of Mexican free-tailed bats *(T. brasiliensis)* collected in Texas [BURNS *et al.,* 1957] and subsequently from the same species collected in California [JOHNSON, 1962]. In the southwestern United States the incidence of natural infection with Rio Bravo virus was shown in a survey of over 1,000 Mexican free-tailed bats to vary from 0 to 7% in groups of 50–275 bats captured at various locations during the summer and fall months, with an overall infection rate of about 2% [CONSTANTINE and WOODALL, 1964]. In studies with *Tadarida* held captive in the laboratory for 1–2 years, Rio Bravo virus was recovered from saliva at intervals throughout the observation period, indicating a carrier state in the naturally infected bat [CONSTANTINE and WOODALL, 1964; BAER and WOODALL, 1966]. As indicated above, during the course of our field studies designed to evaluate the role of Mexican free-tailed bats in the ecology of SLE virus in south Texas, several isolates of Rio Bravo virus were obtained in addition to the strains of SLE virus isolated from this bat species [SULKIN *et al.,* 1966c; ALLEN *et al.,* 1970]. While these field studies were in progress we carried out a limited study on the susceptibility of bats to experimental infection with Rio Bravo virus [SULKIN and ALLEN, unpublished observations]. In these experiments, two bat species *(T. brasiliensis* and *E. fuscus)* were inoculated subcutaneously with about 1,000 WMICLD$_{50}$ of a strain of Rio Bravo virus isolated from a naturally infected Mexican free-tailed bat by Dr. GEORGE BAER. Groups of bats were sacrificed 3, 7, 12 and 18 days postinoculation and blood, brown fat, brain, salivary gland and spleen tissues

were obtained and assayed for virus by the intracerebral inoculation of weanling mice. Rio Bravo virus was demonstrated in one or more of the tissues obtained from bats of both species at each interval tested. Virus was not detected in the salivary glands with any greater frequency than in the other tissues tested. This is of interest since in the survey by CONSTANTINE and WOODALL [1966] brains, lungs, kidneys, mammary glands and fetuses, in addition to salivary glands, were assayed and, with one exception (lung), Rio Bravo virus was recovered only from salivary glands. One might consider this to be evidence that Rio Bravo virus infection is more widespread in the experimentally infected bat than in naturally infected animals; however, Rio Bravo virus has been isolated from the blood and brown fat of naturally infected bats by other investigators [BAER and WOODALL, 1966; SULKIN et al., 1966c; ALLEN et al., 1970].

Entebbe bat salivary gland virus was originally isolated from pools of salivary gland tissue from T. hindei collected in Entebbe, Uganda, East Africa [LUMSDEN et al., 1961]. Subsequently, HI antibodies against this virus were demonstrated in several species of bats in Uganda [SHEPHERD and WILLIAMS, 1964; SIMPSON et al., 1968]. The susceptibility of Rousettus aegypticus to experimental infection with Entebbe bat virus has also been reported [SIMPSON and O'SULLIVAN, 1968].

Dakar bat virus was isolated from the salivary glands of Scotophilus bats collected in Dakar, Senegal [BRÉS and CHAMBON, 1963]. In subsequent field studies more than 3,000 insectivorous bats of several different species were examined; 42 strains of Dakar bat virus were obtained from the assay of 142 pools of salivary gland tissue from these animals and HI antibodies against this agent were demonstrated in 54% of 165 bat serum samples which showed little reactivity with other group B antigens [BRÉS and CHAMBON, 1964]. Strains of Dakar bat virus have also been isolated from Tadarida in Uganda, East Africa [WILLIAMS et al., 1964] and in Nigeria [KARABATSOS, 1969]. Although many tissues from naturally infected bats have been tested, Dakar bat virus has only been isolated from salivary and parotid gland tissue and from saliva. Some evidence of human infection with this agent was also obtained in Senegal; 9 of 303 human serum samples showed exclusive presence of Dakar bat virus antibodies [BRÉS and CHAMBON, 1964].

Montana Myotis leucoencephalitis (MML) virus was recovered by bite transmission to suckling mice from a naturally infected M. lucifugus collected in Montana in 1958 [BELL and THOMAS, 1964]. Subsequent isolations of this agent were made from saliva, salivary gland, brain and brown fat tissues of 3 of 4 M. lucifugus collected in August 1960, and from 10 of 105 bats of the same species collected in 1962. Some evidence was obtained that MML virus may be fatal for *M. lucifugus* and *E. fuscus* when inoculated intracerebrally [BELL and THOMAS, 1964]. No serologic evidence of human infection with MML virus was detected in serum samples from 26 persons residing in the area where bats yielding the agent were collected.

Additional viruses isolated from the salivary glands of naturally infected bats and exhibiting antigenic relationship with group B arboviruses include Bukalasa bat virus from *Tadarida* collected in East Africa [WILLIAMS *et al.*, 1964; FUKUMI *et al.*, 1967], Banqui bat virus from bats collected near Banqui, Central African Republic [CHIPPAUX and CHIPPAUX-HYPPOLITE, 1965], and an agent designated Mount Suswa bat virus by HENDERSON *et al.* [1968], who isolated the virus from the salivary glands of *Otomops martiensseni* collected at Mount Suswa, Kenya, East Africa.

The inter-relationships of certain of the group B bat viruses (Rio Bravo, Entebbe, Dakar, MML and Bukalasa) have been studied by KARABATSOS [1969] using N, HI and CF tests. Although the prototype strains of these viruses are distinguishable from each other, Bukalasa bat virus was shown to be closely related to the Dakar bat virus strains and Rio Bravo virus appeared to be distantly related to MML virus, which by certain tests was also distantly related to Dakar bat virus strains. An earlier study using only the HI test had indicated that Entebbe bat virus might be closely related to MML virus [WILLIAMS *et al.*, 1964]; the results obtained by KARABATSOS [1969] did not support this observation.

The fact that there is no evidence of arthropod transmission for any of the group B bat viruses suggests that these agents may have evolved from true arthropod-borne group B viruses and, in adaptation to the bat, with particular affinity for salivary gland tissue, developed a means of persistence in nature in the absence of an arthropod vector. Laboratory infections with Rio Bravo virus through presumably aerogenic transmission [SULKIN *et al.*, 1962] and bite transmission of MML virus, as demonstrated in the original isolation of that agent [BELL and THOMAS, 1964], indicate means by which these viruses could persist in bat populations in nature. The relationships of the group B bat viruses to other established group B arboviruses not originally isolated from bats have not been studied in depth. However, in the original identification of Rio Bravo virus, the close antigenic relationship of this agent to the SLE virus complex by CF and HI tests was noted [BURNS *et al.*, 1957; JOHNSON, 1962]. There is some controversy in the literature as to the degree of cross-reactivity between these two agents in the serum neutralization test. Certain investigators have shown a one-way cross between SLE and Rio Bravo viruses in which anti-SLE virus serum neutralizes Rio Bravo virus but anti-Rio Bravo virus serum does not neutralize SLE virus [BURNS et al., 1957; SULKIN et al., 1966c]. In a study describing the identification of a strain of SLE virus isolated from a fox, EMMONS and LENNETTE [1967] found little or no cross-reactivity between these two arboviruses and suggested that the demonstrable relationship between the two may depend on the virus strains studied and the test system used. In our studies on the cross-reactivity between SLE and Rio Bravo viruses, we have found that the relationship between these two viruses does, indeed, vary when different strains are compared in different test systems [SULKIN and ALLEN, unpublished observations].

The relationship of Rio Bravo bat virus to Modoc virus, a group B arbovirus isolated from the lactating mammary glands of a *Peromyscus* mouse in California, has been discussed by JOHNSON [1967]. Mice immunized by intramuscular inoculation of Modoc virus were completely immune to intracerebral challenge 1 month later with Rio Bravo virus. Mice immunized by intramuscular inoculation with Rio Bravo virus had only a slight evidence of immunity to Modoc virus. JOHNSON [1967] suggested that Modoc virus has the wider antigenic coverage and that Rio Bravo virus probably derived from Modoc virus; he proposed the name, Modoc-Rio Bravo complex, for this group of viruses.

Yellow fever virus. Early investigators, attempting to show that bats could be involved in the epidemiology of yellow fever, were unsuccessful in demonstrating the transmission of yellow fever virus from infected *Aedes aegypti* mosquitoes to *Molossus* or *Desmodus* bats or the susceptibility of a variety of other bat species to injection of the virus [KUMM, 1932; HUGHES and PERLOWAGORA, 1948]. RODHAIN [1936], however, reported the susceptibility of an African fruit-eating bat *(Epomophorous wahlbergi)* to intracerebral inoculation with a high mouse passage strain of yellow fever virus. Since these early workers were unable to maintain bats in the laboratory for extended periods of time and because it was shown subsequently that bat species may vary in susceptibility to different strains of an arbovirus [SULKIN et al., 1963], we included in our studies on the role of bats in the epidemiology of the arboviruses a limited number of experiments to reinvestigate the susceptibility of these animals to yellow fever virus. In the initial experiment, big brown bats *(E. fuscus)* received about 1,000 WMICLD$_{50}$ of the 17D strain of yellow fever virus; virus was detected in the blood, brown fat, spleen and brain tissue of one adult bat sacrificed 18 days after inoculation, and in only the brown fat tissue of two young bats sacrificed 4 and 7 days, respectively, after receiving virus [SULKIN, 1962]. Additional experiments determined the susceptibility of Mexican free-tailed bats *(T. brasiliensis)* to the French neurotropic strain of yellow fever virus. *Tadarida* proved to be more susceptible to this strain of yellow fever virus than were *Eptesicus* to the 17D strain. Virus was demonstrated in the blood, brown fat, brain, kidney and spleen tissue of bats sacrificed beginning 7 days after subcutaneous inoculation and was shown to persist in the blood of some bats for at least 14 days; at no time were signs of encephalitis observed in any of the animals [SULKIN and ALLEN, unpublished observations]. In more recent studies by other investigators, viremia has been demonstrated in several bat species inoculated intraperitoneally with the French neurotropic strain of yellow fever virus but not with the Seganga strain [SIMPSON and O'SULLIVAN, 1968].

Neutralizing and HI antibodies against yellow fever virus have been demonstrated in various species of bats in Brazil [LAEMMERT *et al.,* 1946], Colombia [PRIAS-LANDINEZ, 1966] and East Africa [SHEPHERD and WILLIAMS, 1964; WILLIAMS *et al.,* 1964; SIMPSON *et al.,* 1968]. Yellow fever virus was isolated from the brain of a bat *(Epomophorous)* collected in Ethiopia during the course of surveys of the sylvatic fauna in that country for evidence of infection with this agent [ANDRAL *et al.,* 1968; SERIE *et al.,* 1968].

Tickborne encephalitis virus (TBE). Serological evidence of natural TBE virus infection in bats was obtained by HAVLIK and KOLMAN [1957] who demonstrated neutralizing antibodies against this agent in serum samples from 4 bat species in Central Bohemia. In subsequent experimental studies it was found that (1) TBE virus infection could be established in bats, (2) the virus could be isolated from blood, brain and liver of infected bats during periods of hibernation, and (3) the virus multiplied in these tissues to even higher levels upon transfer to a warmer environment [NOSEK *et al.,* 1961]. Additional experimental studies with *Myotis* showed that TBE virus was infective for these animals by subcutaneous and oral inoculation and could be recovered from blood, brain, brown fat, liver, spleen and occasionally from fecal specimens [KOLMAN *et al.,* 1960]. In the latter study a few animals exhibited signs of encephalitis and sections of brain tissue showed histological alterations.

Kyasanur Forest disease (KFD) virus. The demonstration of N and HI antibodies against KFD virus in serum samples from the frugivorous bat

R. leschenaulti collected near Poona, India, was reported by PAVRI and SINGH [1965]. In a preliminary study involving a small number of bats, these investigators demonstrated the susceptibility of this species to experimental KFD virus infection as evidenced by detection of a low-grade viremia followed by the development of N and HI antibodies in all inoculated bats. Subsequently, N antibodies were demonstrated in 3 additional bat species, *Rhinolophus rouxi, Cynopterus sphinx* and *Pteropus giganteus* [RAO, 1968; RAJAGOPALAN *et al.,* 1969]. During 1968, the Virus Research Center at Poona carried out large-scale field studies to define the role of bats in the epidemiology of KFD. Neutralizing antibodies against KFD virus were demonstrated in 32 of 102 serum samples from *R. leschenaulti* and in 19 of 453 serum samples from *R. rouxi,* all members of a single colony of bats inhabiting an abandoned well at Kasarguppe, within the KFD area [RAO, 1968; RAJAGOPALAN *et al.,* 1969]. A small number of larvae of *Ornithodoros* ticks taken off the bats were tested for KFD virus with negative results; however, KFD virus was isolated from a pool of 5 adult *Ornithodoros* collected from the same well inhabited by the colony of *R. rouxi* [RAJAGOPALAN *et al.,* 1969]. In studies of experimental infection of *C. sphinx* with KFD virus it has been shown that following intramuscular inoculation virus can be demonstrated in blood, brain, salivary gland, heart, lung, spleen, liver and kidney tissue; N antibodies were detected in serum samples from bats which survived for 21 days [PAVRI and SINGH, 1968]. In contrast to *R. leschenaulti,* which showed no signs of illness due to experimental infection with KFD virus, the majority of *C. sphinx* succumbed to the infection.

West Nile virus. Serological evidence of infection with West Nile virus has been reported for bats in Egypt [TAYLOR *et al.,* 1956], Australia [ROWAN, 1957; ROWAN and O'CONNOR, 1957], East Africa [SHEPHERD and WILLIAMS, 1964; SIMPSON *et al.,* 1968], Italy [BALDUCCI, 1968], Ethiopia [ANDRAL *et al.,* 1968], and Israel [AKOV and GOLDWASSER, 1966]. West Nile virus has been isolated from the spleen tissue of a bat *(R. leschenaulti)* collected in the Kyasanur Forest disease area of India [PAUL *et al.,* 1970].

Only limited data are available concerning the association of bats with other group B arboviruses listed in tables V and VI (Zika, Dengue, Ntaya, Usutu, Ilheus, Murray Valley encephalitis, Tembusu, Wesselsbron, Bussuquara, Uganda S and Turkey meningoencephalitis viruses). Still other group B viruses which have been recovered from bats remain to be characterized [CAREY *et al.,* 1968; SALAUN, 1970; SULKIN *et al.,* 1970].

Association of Bats with Group A and other Arboviruses

Evidence of natural and experimental infection of bats with group A arboviruses is much less extensive than that accumulated for the group B arboviruses. Chikungunya virus has been isolated from the salivary glands of bats *(Scotophilus)* collected in Senegal [BRÉS *et al.,* 1963] and the presence of HI antibodies against this agent in serum samples from bats collected in Senegal and Uganda has also been reported [BRÉS and CHAMBON, 1964; SHEPHERD and WILLIAMS, 1964]. Experimental infection of *P. giganteus* with Chikungunya virus was accomplished by SHAH and DANIEL [1966], who demonstrated viremia followed by the development of N and HI antibodies in sera from infected bats. SHEPHERD and WILLIAMS [1964] detected HI antibodies against Semliki Forest (SF) virus in serum samples from bats collected in East Africa and experimental infection of 3 bat species *(Eidolon helvum, R. aegypticus* and *T. condylura)* with this agent was reported by SIMPSON and O'SULLIVAN [1968], who demonstrated SF virus in blood, brain, salivary glands and kidneys of infected bats that showed no signs of illness. These same investigators infected *E. helvum* with Sindbis virus, as evidenced by the detection of virus in the blood and subsequent demonstration of N antibodies [SIMPSON and O'SULLIVAN, 1968]. The isolation of WEE virus from the salivary glands of an *E. fuscus* collected in New Jersey was reported to us by GOLDFIELD [1968]. CONSTANTINE [1970] cites serological evidence of natural infection of bats with this agent in Brazil and also of experimental WEE virus infection of *T. brasiliensis* by an investigator in the United States; in the latter study mosquitoes were infected by feeding on the viremic *Tadarida.*

Several studies have been concerned with the association of bats with EEE virus in the United States. Neutralizing antibodies against EEE virus have been demonstrated in serum samples from bats collected in Georgia [KARSTAD and HANSON, 1958] and Massachusetts [DANIELS *et al.,* 1960; MAIN, 1970] and this virus has been isolated from brains, salivary glands and liver-spleen pools from bats collected in New Jersey and Pennsylvania [GOLDFIELD, 1968]. LAMOTTE [1958], in a report concerning experimental JBE virus infection in *E. fuscus,* noted viremia without signs of illness in bats of this species inoculated with EEE virus. In more recent studies it was shown that in *E. fuscus, M. keenii,* and *M. lucifugus* EEE virus produced a prolonged viremia with virus also demonstrable in mammary gland, brown fat, pancreas, lungs, kidneys and liver and, less frequently, in brain, salivary glands and heart; no signs of encephalitis were noted in any of the infected bats [MAIN, 1970].

One of the earliest studies of experimental arbovirus infection in bats was concerned with Venezuelan equine encephalitis (VEE) virus in *E. fuscus M. lucifugus, Plecotus townsendii* and *P. subflavus.* CORRISTAN *et al.* [1956, 1958] reported persistent infection of these bat species with VEE virus characterized by high-level viremias over a period of 2–26 days with no overt signs of illness. Infected bats sustained infection during a prolonged period of dormancy at 10 °C and upon transfer to a warmer environment experienced levels of viremia sufficient to infect mosquitoes. Evidence of natural VEE virus infection in bats in Colombia was obtained by PRIAS-LANDINEZ [1966], who isolated the virus from organ pools of *Carollia perspicillata;* HI antibodies against this agent were demonstrated in serum samples from *Artibeus* collected in Panama [GRAYSON and GALINDO, 1968]. SCHERER *et al.* [1971], during studies on the ecology of VEE virus in southeastern Mexico, isolated VEE from an *Artibeus turpis* collected in 1963 and from a *C. subrufa* collected in 1966; a total of 614 bats collected during July and August 1963–1965 and April 1966 were tested for evidence of VEE virus infection in these studies. Also in these studies, N and HI antibodies against VEE virus were demonstrated in sera from 3 bat genera *(Artibeus, Carollia* and *Glossophaga).* During an equine epizootic of VEE in the southeastern Mexican states of Chiapas and Oaxaca in 1970, 18 bats were collected and examined for evidence of infection with VEE virus [CORREA-GIRON *et al., 1972*]. Bat species examined included 6 *Balantiopteryx plicata,* 8 *Mollosus* species and 4 vampire bats *(Desmodus rotundus),* and tissues removed for assay included blood, brown fat, brain, submaxillary salivary glands and viscera (lung, heart, kidney, liver and spleen). VEE virus was recovered from a 10-percent suspension of the viscera of 1 of the 4 vampire bats. Other tissues from this bat were negative for VEE virus as were all tissues of the other 17 bats tested. Neutralizing antibodies against VEE virus were detected in the serum from the bat whose viscera yielded VEE virus but not in serum samples from the uninfected animals. This vampire bat strain of VEE virus was shown by kinetic HI studies to be closely related to an epidemic strain of VEE virus, type 1B. We obtained the vampire bat strain of VEE virus from the National Center for Disease Control in Atlanta, Georgia, and studied the characteristics of infection induced by this strain in an insectivorous bat species *(E. fuscus).* Following intraperitoneal inoculation of a small dose (about 200 weanling mouse intraperitoneal LD_{50}), viremia was demonstrable in 100% of the bats and within 48–72 h VEE virus was detected in brain, brown fat and spleen tissue; viremia subsided within 10–13 days as N antibodies appeared in the circulation, but virus persisted in other tissues of some bats for at least 3 weeks in the absence of overt signs of encephalitis [SULKIN and ALLEN, unpublished observations].

In addition to the evidence of the association of bats with group B and group A arboviruses listed in tables V and VI as discussed above, natural and experimental infection of bats with Bunyamwera virus has been reported [SHEPHERD and WILLIAMS, 1964; WILLIAMS et al., 1964; SIMPSON et al., 1968; SIMPSON and O'SULLIVAN, 1968]. Also, CONSTANTINE [1970] cites information accumulated through personal communications concerning natural infection of bats with a number of miscellaneous arboviruses.

Natural and Experimental Infection of Bats with Miscellaneous Viruses

Table XIV presents evidence of natural and experimental infection of bats with miscellaneous viruses. The rhabdovirus group includes 3 bat viruses (Kern Canyon [KC], Lagos bat [LB] and Mount Elgon bat [MEB]) which were initially given provisional classification as arboviruses. KC virus was isolated from a pool of spleen-heart tissue from a bat *(M. yumanensis)* collected in Kern Canyon, California, in 1956; a second isolation from the same species was made in 1961 [JOHNSON, 1965]. This agent was tentatively considered an arbovirus on the basis of similarities of some biological properties with group B arboviruses isolated from bats until justification for its inclusion within the rhabdovirus family of viruses was obtained by MURPHY and FIELDS [1967] based on electron microscopic and immunological studies. LB virus was isolated from a pool of brain tissue from 6 Nigerian fruit bats *(E. helvum)* in 1956 by BOULGER and PORTERFIELD [1958] and was subsequently found, along with another virus isolated from shrews in Nigeria, to be serologically and morphologically related to rabies virus by SHOPE *et al.* [1970] who suggested that these agents, together with rabies, form a subgrouping within the rhabdoviruses. MEB virus was isolated from the salivary glands of an insectivorous bat *(R. hildebrandtii)* in Kenya, Africa, in 1964 [METSELAAR *et al.*, 1969]. In the original identification it was found that the agent was not neutralized by rabies antiserum, and CF tests against a variety of arboviruses failed to identify it or indicate any antigenic relatedness to any virus known to occur in Africa. Multiplication of MEB virus in *A. aegypti* following inoculation was demonstrated but infection by feeding was not achieved and attempted transmission experiments were negative. The agent was placed provisionally among the ungrouped arboviruses, however, because of certain similarities with other agents in this category which had been isolated from bats. Subsequent structural and immunological determination provided a basis for considering MEB virus an antigenically distinct member of the rhabdovirus group [MURPHY *et al.*, 1970].

Although the suggestion that bats might be involved in the natural history of vesicular stomatitis was made by HANSON about 20 years ago

Table XIV. Natural and experimental infection of bats with miscellaneous viruses

Virus	Bat genera	Natural infections		Experimental infections	References
		country	evidence		
Rhabdoviruses					
Kern Canyon	*Myotis*	United States	Sp-He pool[1]		JOHNSON, 1965
Lagos bat	*Eidolon*	Nigeria	Br		BOULGER and PORTERFIELD, 1958
Mt. Elgon bat	*Rhinolophus*	Kenya	SG		METSELAAR *et al.*, 1969
Vesicular stomatitis	*Myotis*			Bl	DONALDSON, 1970
	Carollia, Artibeus,	Panama	N antibody		TESH *et al.*, 1969
	Glossophaga, Tonatia,				
	Lonchophylla				
Arenavirus					
Tacaribe	*Artibeus*	Trinidad	Br, Li, SG, Sp	organ pools	DOWNS *et al.*, 1963
	Desmodus			N antibodies	
Reovirus					
Reovirus, type 1	*Eptesicus*	Australia	HI antibodies		STANLEY *et al.*, 1964
Picornaviruses					
Coxsackievirus, B3	*Eptesicus*			Bl, Br, BF, He	DEMPSTER *et al.*, 1961
Poliovirus, type 2	*Myotis*			Br	REAGAN *et al.*, 1954
Paramyxovirus					
Parainfluenza, type 2	*Rousettus*	India	Br-SG-Sp pools		PAVRI *et al.*, 1971
			N, HI antibodies		HOLLINGER and PAVRI, 1971

1 Tissue abbreviations: Bl = blood; Br = brain; BF = brown fat; He = heart; Li = liver; SG = salivary gland; Sp = spleen.

[HANSON, 1952] no evidence to support this hypothesis appeared in the literature until 1969 when ecologic studies of vesicular stomatitis virus (VSV) in Panama included several species of bats among animals tested for neutralizing antibodies against VSV [TESH et al., 1969]. Both VSV-Indiana and VSV-New Jersey antibodies were detected in several species of bats. In fact, bats were among the 3 animal populations with the highest infection rates against VSV-New Jersey. The susceptibility of little brown bats (M. lucifugus) to experimental infection with the Cocal serotype of VSV has been studied recently [DONALDSON, 1970]. Animals infected by subcutaneous inoculation circulated virus for 10–16 days, and transmission of the virus from infected bats to suckling mice by A. aegypti was demonstrated.

Several strains of Tacaribe (TCR) virus were isolated from two species of American fruit bats (A. jamaicensis and A. literatus) collected in Trinidad during 1956–1958 [DOWNS et al., 1963]. A single isolation of the same agent was made from a pool of 344 mosquitoes, representing 8 genera and 19 species, collected in Trinidad in 1956. Support of this mosquito isolation was not obtained in an interval from 1956 to 1962 although over 1,000,000 mosquitoes were processed during that time. TCR virus was found to be immunologically distinct from numerous arboviruses, LB virus, and rabies virus but was placed with the ungrouped arboviruses based on its inactivation by sodium desoxycholate; it has now been placed, along with other members of the Tacaribe complex, in a new distinct taxonomic group – the arenaviruses – a name chosen to reflect the fine granules seen in the virion in ultrathin sections. Studies on experimental TCR virus infection in A. jamaicensis, A. literatus and D. rotundus showed that these species are not very susceptible to intramuscular inoculation. Viremia could not be demonstrated and no apparent illness developed, although virus was recovered from the internal organs in an occasional bat [DOWNS et al., 1963].

In studies on the ecology and epidemiology of reovirus in Australia, serum samples from 80 bats (E. pumilus) were collected and made into 3 pools for testing; no HI antibodies against reovirus types 2 and 3 were detected but all 3 pools contained HI antibodies against type 1 reovirus [STANLEY et al., 1964].

Big brown bats (E. fuscus) have been shown to be susceptible to experimental infection with coxsackie B3 virus. Following subcutaneous inoculation, virus was demonstrable in blood, brown fat and heart tissue of bats maintained at 20–22 °C for 4 days and in blood, brown fat, brain and heart tissue of bats maintained in a hibernating state at 2 °C for 140 days [DEMPSTER et al., 1961]. REAGAN et al. [1954] reported that M. lucifugus were slightly susceptible to the Lansing strain of poliovirus by intracerebral or oral inoculation or the mouse-adapted MEF1 strain of poliovirus by intracerebral and intrarectal inoculation as demonstrated by subinoculation of bat brain tissue into mice. In limited experiments in our laboratory we were unable to infect *E. fuscus* by intracerebral or oral inoculation of the mouse-adapted MEF1 strain of poliovirus nor were bat embryo cell cultures susceptible to this agent [SULKIN and ALLEN, unpublished observations].

Finally, a paramyxovirus, identified as a new animal subtype of the parainfluenza type 2 group, serologically related to, but distinct from SV41 virus, was isolated from fructivorous bats *(R. leschenaulti)* collected at Poona, India [PAVRI *et al.,* 1971; HOLLINGER and PAVRI, 1971]. HI antibodies against the bat parainfluenza type 2 virus were demonstrated in 5 of 70 *Rousettus* and in 20 of 200 human serum samples tested.

Studies of Factors Relating to the Inapparency of Virus Infections in Bats

During the course of studies in our laboratory, designed primarily to demonstrate experimentally how the bat could serve as a reservoir host in nature for rabies virus and for certain arboviruses [SULKIN *et al.*, 1959, 1960, 1963, 1964, 1966a, b], it became apparent that these animals were unique in their response to virus infections. Although all strains of viruses tested were capable of replicating in various tissues of the experimentally infected bat (including brain tissue), only in experiments with one strain of rabies virus were signs of central nervous system disease apparent in a small percentage of animals [SULKIN *et al.*, 1959]. Experimental JBE and SLE virus infections in bats proved to be classic examples of inapparent virus infections [SULKIN *et al.*, 1963]. Aside from demonstrating how bats could serve as reservoir hosts for rabies virus and arboviruses in nature, these studies suggested that experimental virus infections in bats would provide excellent models for the study of virus-host cell interactions which do not result in cell damage leading to overt disease or death. Thus, studies of factors relating to the inapparency of virus infections in bats were included in our research program. We have already discussed experiments with gravid bats which demonstrated that the physiological stress of pregnancy did not alter the course of rabies virus infection in *T. brasiliensis* [SIMS *et al.*, 1963] and that gravid *Tadarida* were not more susceptible to JBE or SLE viruses than were non-gravid bats of the same species [SULKIN *et al.*, 1964]. Also, in studies on the influence of environmental temperature on arbovirus infections in bats we found that the increased body temperature and metabolic rate resulting from the maintenance of bats infected with JBE virus at 37 °C did not induce signs of encephalitis or apparent damage to brain tissue [SULKIN *et al.*, 1966a].

The Effects of 6-Thioguanine and Endotoxin on Experimental JBE and SLE Virus Infections in Bats

Another approach to defining the factors which enable bats to sustain virus infections in the absence of overt disease was an investigation of the influence of certain substances, known to affect the susceptibility or resistance of other animal species to virus infections, on the course and outcome of virus infections in bats. We had already determined in experiments correlated with our studies with gravid bats that administration of cortisone, a substance which had been shown to influence host susceptibility to many experimental virus infections [KILBOURNE and HORSFALL, 1951; SULKIN et al., 1952; SHWARTZMAN, 1953; BRYGOO and DODIN, 1957; IMAM and HAMMON, 1957; BUGBEE et al., 1960], did not increase the susceptibility of T. brasiliensis to a strain of SLE virus of low infectivity for this bat species [SULKIN et al., 1964]. Since bacterial endotoxin had been shown under various circumstances to alter the susceptibility of mice to certain bacterial and viral infections [BOEHME, 1960; GLEDHILL, 1959; GOODMAN and KOPROWSKI, 1962] and 6-thioguanine, a radiomimetic compound, was known to suppress resistance of mice to virus infection [GOODMAN and KOPROWSKI, 1962], the influence of these two substances on the course of experimental JBE and SLE virus infection in two species of bats (M. lucifugus and T. brasiliensis) was determined [MIDDLEBROOKS et al., 1968]. The results showed that treatment with either substance enhanced the susceptibility of both bat species to JBE and SLE virus infection as evidenced by higher infection rates and increased levels of virus in brain, lung, spleen and liver tissue from treated bats as compared to control animals sacrificed at the same time following virus inoculation. Although treatment with endotoxin or thioguanine increased the multiplication of JBE and SLE virus in various tissues of the bats, no overt signs of encephalitis were observed in any of the treated animals during the course of the experiments. In correlated studies designed to detect the primary sites of localization and replication of JBE and SLE viruses in bats, it was found that the liver rapidly removed injected virus from the blood and that this organ was a prime site for virus multiplication; in all cases virus apparently multiplied originally in the vascular endothelium and macrophages [MIDDLEBROOKS, 1966]. Since both 6-thioguanine and endotoxin are believed to influence susceptibility of animals to bacterial and viral infections through action on the reticuloendothelial (RE) system, it may be inferred that the success of viral infections in the bat, as in the mouse, may depend to a great extent on the response of the RE system to the infecting virus. However, even though bats treated with thioguanine and endotoxin suffer more intense and widespread infections with JBE and SLE viruses, the inapparent infection is not converted to overt disease and death by the action of these substances.

Investigations of the Immune Capabilities of Bats

Previous studies on the immune response of bats experimentally infected with JBE virus had shown that these animals did not produce CF or HI antibodies against this agent; neutralizing antibodies could be detected in the plasma of most, but not all, infected bats and, when formed, was not always long-lasting or protective [SULKIN et al., 1966c]. Investigations of the immune capabilities of these animals were extended to obtain data which could be used in evaluating the effect of antibody formation on the course and outcome of viral infections in bats. In experiments designed to characterize the immunoglobulins formed by bats infected with JBE virus, it was shown that infected bats were capable of producing both heavy (19S) and light (7S) immunoglobulins [LEONARD et al., 1968]. The association of CF activity with 7S immunoglobulins has been reported by several investigators [PIKE, 1967]. Thus, it would appear that failure to detect CF antibodies in bats experimentally infected with JBE virus is not due to the failure of these animals to form a low molecular weight immunoglobulin. However, this particular immunoglobulin may lack the structural conformation (Fc fragment) to react with complement. In comparing the neutralizing capacity of hyperimmune bat plasma prepared against JBE virus with that of hyperimmune guinea pig serum against this agent we found that guinea pigs produced 100-fold higher titers of neutralizing antibodies than did bats; however, in a comparison of hyperimmune bat plasma and hyperimmune guinea pig serum against JBE virus by a serum-dose/virus-response method [RUSSELL et al., 1967], the N slopes obtained were essentially the same [LEONARD et al., 1968]. These results indicate that the hyperimmune bat plasma and hyperimmune guinea pig serum had the same specificity, reacted with identical receptors of the virus and that equal numbers of bat or guinea pig antibodies are required to neutralize the infectivity of JBE virus. Thus, there is possibly only a quantitative deficiency of the total immune response of bats infected with JBE virus rather than qualitative deficiencies in the immunoglobulins produced.

Because much definitive information was available in the literature concerning the immune response of laboratory animals to an *Escherichia coli* bacteriophage ∅X174 [UHR and FINKELSTEIN, 1963], experiments were devised to compare the development, nature and neutralizing capacity of bat antibodies formed in response to this antigen to that observed in conventional laboratory animals. The immune response of bats, guinea pigs, and rabbits to a series of 3 primary injections and a booster injection of bacteriophage ∅X174 was studied [HATTEN *et al.*, 1968]. Anti-phage activity developed more rapidly in bats maintained at 24 °C and 37 °C than in other animals, but the magnitude and duration of the primary response was less, particularly in 24 °C bats. Prolonged primary production of mercaptoethanol (ME)-sensitive antibody was observed in 24 °C bats, and 37 °C bats had a large proportion of ME-sensitive antibody throughout the first 21 days. Kinetic studies of ∅X174 inactivation indicated that 37 °C bat antibodies, developed during the latter part of the primary response, did not bind phage irreversibly. A booster injection of ∅X174 enhanced the neutralizing capacity of antibodies produced in all 3 animal species. Approximately 40% of the anti-phage activity in 10-day postbooster plasma pools from 24 °C bats were due to ME-sensitive antibodies. Guinea pigs, rabbits and 37 °C bats had little detectable ME-sensitive antibodies during the secondary response by the assay procedures used. While the rate and extent of phage inactivation by a 30-day postbooster plasma pool from 37 °C bats greatly exceeded that obtained with a comparable pool from 24 °C bats, bat antibodies did not inactivate ∅X174 as rapidly or to the same degree as 30-day postbooster guinea pig or rabbit antisera. From these results it was concluded that the primary and secondary immune response of bats to immunization with ∅X174 was qualitatively as well as quantitatively less than the response of rabbits and guinea pigs to the same antigen. Confirmation for this conclusion was obtained in subsequent studies in which K values, based on 90% neutralization of ∅X174, were obtained which indicated that the response of bats held at 24 and 37 °C to multiple doses of 10^{10} plaque-forming units of the bacteriophage was less and that production of ME-sensitive antibodies was more pronounced than in rabbits or guinea pigs [HATTEN *et al.*, 1970]. Also, considerable dissociation of ∅X174 from bat antibodies occurred when less than 90% of the phage had been neutralized, but was not remarkable when inactivation was greater than 90%. The proportionate loss of anti-∅X174 activity in relation to concentration of antibodies diminished upon dilution of bat plasma pools as the immune response progressed, suggesting that an increase in affinity of antibodies for ∅X174 had occurred, possibly in association with changes in the major type of antibodies being produced.

In general, results of studies on the immune response of bats to a variety of antigens tend to agree that their response is different from homoiothermic animals, although perhaps no different than other hibernating mammals [MCKENNA and MUSACCHIA, 1968; JAROSLOW, 1968]. The quantity and quality of antibodies synthesized by bats may be especially dependent upon the degree and nature of the antigenic stimulation since optimal conditions for maximum antibody production are more critical than in homoiothermic animals where metabolic processes are more stable. It has been proposed that more avid antibodies are bound by excess antigen *in vivo* and as a result are not detected by *in vitro* assays [GOOD and MOLLER, 1969; GOIDL et al., 1968]. Bats are known to harbor infectious viruses for long periods of time [SULKIN, 1962; SULKIN et al., 1966c; BAER and WOODALL, 1966; BELL et al., 1969] and in the studies with ∅X174 there were indications that phage may have persisted *in vivo* long after the final doses were administered [HATTEN et al., 1970]. Thus, apparent deficiencies in the immune response of bats, particularly with regard to the quantity and quality of circulating antibodies, are not conclusive evidence that a basic anomaly exists in the immune capabilities of this species.

In another approach to defining the role of the immune response in persistent virus infections in bats we studied the effect of an immunosuppressant drug, cyclophosphamide (Cytoxan), on JBE virus infection in bats [SULKIN and ALLEN, unpublished observations]. We found that Cytoxan does suppress antibody formation in JBE virus-infected bats and that some treated bats suffered more intense and prolonged viremias than untreated animals; however, even those bats experiencing extended periods of viremia showed no overt signs of disease. Also, in some immunosuppressed animals there was a cessation of the viremic state within 10–14 days after virus inoculation, suggesting control of infection by mechanisms other than the immune response.

The Role of Interferon in Inapparent Virus Infections in Bats

Studies undertaken to investigate the role of interferon in the ability of bats to sustain chronic virus infections have shown that bats and bat cell cultures that are persistently infected with JBE virus do produce interferon [STEWART et al., 1969; STEWART, 1970]. In the intact animal interferon was produced by brain, brown fat and spleen but not by kidney, liver or lung tissue. In the brain, virus levels were paralleled during the entire course of the infection by the levels of interferon measured. Since JBE virus was shown to be very sensitive to bat interferon [STEWART *et al.*, 1969], it is tempting to speculate that the interferon produced in this tissue kept the virus production from attaining maximum levels, thereby preventing pathology which possibly would result if virus concentrations in brain tissue had reached higher levels. However, if interferon were responsible for the suppression of virus multiplication in bats, one would expect virus to be eliminated from the animal. The apparent persistence of foci of JBE virus infection in chronically infected bats which give rise to recurrent cycles of virus replication suggests that perhaps certain populations of cells within the animal are not sensitive to the action of interferon.

In *in vitro* studies using lines of bat embryo cells [STEWART *et al.*, 1970] it was found that bat brain passaged JBE virus yielded higher levels of interferon than mouse brain passaged virus [STEWART, 1970]. Cloned bat embryo cells, in which interferon-dependent JBE virus infections could be produced, yielded virus progeny which were mixtures of noninfectious and infectious virions. On the basis of heat and ultraviolet (UV) inactivation, it was shown that untreated virus progeny from bat brain and from bat embryo cells were similar to mixtures of active and UV-inactivated virions of mouse cell origin in interferon inducing ability and in capacity for producing persistent infections *in vitro*. This suggests that the innate capacity of bat brain cells to produce few infectious and many noninfectious, interferon-inducing virus particles may be related to the ability of bats to become persistently infected with arboviruses. In mice, a strain of JBE virus which caused an intense viremia and induced low levels of interferon was more virulent than a strain that did not cause viremia and induced high levels of interferon [ROKUTANDA, 1969].

The Importance of Virus Infections in Bats

The problems caused by the existence of rabies virus infection in vampire bat populations in the Latin American countries are well recognized. Human deaths due to bites of rabid vampire bats continue to occur and, in addition, vampire bat-transmitted rabies is the major cause of death in cattle in these countries, with mortality estimated at upwards of 1,000,000 head each year at a cost of almost $ 50,000,000 [ACHA, 1968]. Over the years, various methods for reducing vampire bat populations (gassing, shooting, dynamiting) have been explored, but mass vaccination of cattle, though not significantly effective, has been the major weapon used against bovine rabies [GREENHALL, 1971]. Recent recognition of the toxicity of anticoagulants for vampire bats and the development of methods for administering these substances to the bats (1) through systemic treatment of cattle with diphenadione, which provides toxic blood meals for feeding vampires [THOMPSON et al., 1972], or (2) by topical application of a jelly containing chlorophacinone to a few netted bats which, when released, spread the toxicant to other bats in the colony by contact and grooming [LINHART et al., 1972], may provide the long searched for effective and practical means for controlling vampire bat populations.

The importance of rabies virus infection in populations of insectivorous bat species native to the United States is not as obvious as is the case for the vampire bats in Latin American countries and investigators continue to characterize and evaluate bat rabies in the southwest [CONSTANTINE, 1967b], the northwest [BARKER and FRANCIS, 1972] and throughout the United States [BAER and ADAMS, 1967]. It would seem clear from data presented in table II that during the past 10 years insectivorous bats of the United States have emerged as a significant wildlife host of rabies virus. It is of related interest to cite data presented by WINKLER [1972] which points up a dramatic shift in the species distribution of animal rabies in the United States between 1951 and 1970. Prior to 1951, reported cases of rabies were predominantly domestic species (dogs and cats); the existence of wildlife rabies (primarily skunks and foxes) was recognized but not considered important since human rabies was associated only with bites of dogs or cats. From 1960 onward, reported cases of rabies in wild animals have exceeded domestic animal cases and during the last 5 years of the transition period (1966–1970) all human rabies cases reported were from exposure to wildlife, with the exception of one possible dog-transmitted case [WINKLER, 1972].

Contact with insectivorous bats has resulted in 7 human deaths from rabies in the United States. The first reported case concerned a woman who died of rabies 25 days after being bitten on the left forearm by a bat which was not identified as to species or examined for evidence of rabies infection [SULKIN and GREVE, 1954]. The second case to occur in the United States was a member of the Entomology Section of the Texas State Department of Health who died of rabies in January 1956 after handling bats in field and laboratory studies; no evidence of a bat bite was recorded [IRONS et al., 1957]. The first instance in which the bite of a rabid bat was unequivocally the cause of a fatal case of human rabies occurred in California in 1958 [HUMPHREY, 1960]. The patient, bitten on the middle finger of the left hand while handling a silver-haired bat *(Lasionycteris noctivagans)* which behaved abnormally, died of rabies following an incubation period of about 57 days and an illness of 8 days. Rabies developed in this patient despite prompt use of combined seroprophylaxis and rabies vaccine of duck embryo origin. The animal was proved rabid by the finding of Negri bodies in impression smears of bat brain material and isolation of the virus by mouse inoculation [LENNETTE et al., 1960]. The precise nature of the exposure of a mining engineer who died of rabies in June 1959 is not clear, although there was evidence that a bat nicked him in the face while he was exploring bat caves in the Big Bend area of Texas for possible guano mining locations [KENT and FINEGOLD, 1960]. In 1959, also, a farmer in Wisconsin died of rabies 28 days after being bitten on the right ear lobe by a bat which was not identified or examined for rabies virus [Vital Statistics of the United States, 1959]. A young boy died of rabies in Idaho in 1960, possibly the victim of a rabid bat, and the most recent death from bat-transmitted rabies occurred in New Jersey in 1971 when a 64-year-old man died 67 days after being bitten on the lip by a bat (National Center for Disease Control, Annual Rabies Summary, 1971). Still another case of bat-transmitted rabies occurred in the United States in 1970, but the victim, a 6-year-old boy who developed clinical rabies 20 days after being bitten and 2 days after completing a 14-day course of duck embryo vaccine, recovered completely; this represents the first adequately documented recovery from rabies in man [HATTWICK et al., 1972].

The extent of the role bats may play in the persistence of rabies in wildlife populations by transmission to other wildlife hosts is difficult to evaluate, but it is known that bats come in contact with other species, especially those which prey on cave bat colonies [CONSTANTINE, 1948]. Studies which have demonstrated oral infectivity of rabies virus [SOAVE, 1966; FISCHMAN and WARD, 1968; CORREA-GIRON, 1970] suggest that this means of interspecies transmission of rabies virus could occur in nature. Also, the demonstration of airborne transmission of rabies virus in bat caves [CONSTANTINE, 1967a] indicates another means through which bats could perpetuate rabies virus in nature as well as endanger man.

An interesting aspect of bat rabies is the recognition that strains of rabies virus which circulate in bat populations may differ from strains isolated from other animal species. Even in the early studies of PAWAN [1936b] it was noted that rabies virus strains from vampire bats were less neurotropic than classical canine strains and it has been suggested that vampire bat strains may be of low infectivity for humans [MALAGA ALBA, 1954]. Rabies virus strains with unusual characteristics have been isolated from bats in Yugoslavia, Turkey and Nigeria [NIKOLIC and JELESIC, 1956; JELESIC and JELESIC, 1959; NIKOLITSCH, 1959] and also in the United States [ENRIGHT, 1955; STAMM et al., 1956]. In our experimental studies we had the opportunity to compare characteristics of a rabies virus strain of canine origin with those of a strain isolated from the brown fat of a naturally infected bat and found significant differences between these two strains, the bat strain being less neurotropic and having a longer incubation period in mice [SULKIN et al., 1960]. The possibility that strain difference was involved in the recovery of the young boy from bat-transmitted rabies was considered, but the strain isolated from the bat that bit him exhibited virulence in mice similar to other street strains of rabies virus [HATTWICK et al., 1972]. The fact that two of the individuals who died of bat-transmitted rabies had received duck embryo vaccine following exposure [HUMPHREY et al., 1960; Vital Statistics of the United States, 1959] suggests that strains of rabies virus of increased virulence or unusual pathogenecity may also circulate in bat populations.

The degree to which bat species throughout the world may contribute to the maintenance and dissemination of the arboviruses would appear to be potentially significant. The field studies conducted in Japan [SULKIN et al., 1970; MIURA et al., 1970] together with those carried out in the United States [ALLEN et al., 1970] established that bats can be natural hosts of JBE and SLE viruses. The isolation of these viruses from bats collected during all seasons of the year indicates that these animals are rare hosts, capable of

maintaining persistent foci of virus activity in endemic areas. The detection
of recurring SLE virus infection at a homesite in Florida over a 2-year period
suggested the existence of urban enzootic foci of this virus [EHRENKRANZ
et al., 1967]. The results obtained in field studies in Houston and Corpus
Christi, Texas [ALLEN et al., 1970] indicate that colonies of Mexican free-
tailed bats could constitute SLE virus-infected sites where the agent could
persist until the convergence of those factors that precipitate epidemics of
SLE. The nocturnal-feeding flight patterns and seasonal migratory movements of bats suggest that they could also be effective in the dissemination
of arboviruses within endemic areas or in the reintroduction of these agents
into certain locales each spring.

Perhaps the most interesting aspect of the study of natural and experimental viral infections in bats is the lack of response of the host to established
infection. Bats may suffer no ill effects from infection with rabies virus and
inapparent infection of bats with certain arboviruses is the rule rather than
the exception. Yet these animals are capable of transmitting rabies by bite
and of providing infective blood meals for vectors of arboviruses, thus ful-
filling the prime requisite for an ideal reservoir host.

References

ACHA, P. N.: Epidemiologia de la rabia bovina paralitica transmitida por los quiropteros. Bol. Ofic. Panamer. *64:* 411–430 (1968).

AKOV, Y. and GOLDWASSER, R.: Prevalence of antibodies to arboviruses in various animals in Israel. Bull. Wld Hlth Org. *34:* 901–909 (1966).

ALLEN, R.; SIMS, R. A., and SULKIN, S. E.: Studies with cultured brown adipose tissue. I. Persistence of rabies virus in bat brown fat. Amer. J. Hyg. *80:* 11–24 (1964a).

ALLEN, R.; SIMS, R. A., and SULKIN, S. E.: Studies with cultured brown adipose tissue. II. Influence of low temperature on rabies virus infection in bat brown fat. Amer. J. Hyg. *80:* 25–32 (1964b).

ALLEN, R.; TAYLOR, S. K., and SULKIN, S. E.: Studies of arthropod-borne virus infections in Chiroptera. VIII. Evidence of natural St. Louis encephalitis virus infection in bats. Amer. J. trop. Med. Hyg. *19:* 851–859 (1970).

ANDRAL, L.; BRÉS, P.; SERIE, C.; CASALS, J. et PANTHIER, R.: Etudes sur la fièvre jaune en Ethiopie. 3. Etude sérologique et virologique de la faune sylvatique. Bull. Wld Hlth Org. *38:* 855–861 (1968).

ARONSON, S. M. and SHWARTZMAN, G.: The histopathology of brown fat in experimental poliomyelitis. Amer. J. Path. *32:* 315–333 (1956).

ARONSON, S. M.; TEODORU, C. V.; ADLER, M., and SHWARTZMAN, G.: Influence of cortisone upon brown fat of hamsters and mice. Proc. Soc. exp. Biol. Med. *85:* 214–218 (1954).

ATANASIU, P. and LEPINE, P.: Multiplication of rabies street virus on mouse ependymoma in tissue culture. Cytolytic effect. Ann. Inst. Pasteur *96:* 72–78 (1959).

AVERY, R. J. and TAILYOUR, J. M.: The isolation of the rabies virus from insectivorous bats in British Columbia. Canad. J. comp. Med. *24:* 143–146 (1960).

BAER, G. M.: Personal commun. (1972).

BAER, G. M. and ADAMS, D. B.: Rabies in insectivorous bats in the United States, 1953–65. Publ. Hlth Rep., Wash. *85:* 637–645 (1970).

BAER, G. M. and BALES, G. L.: Experimental rabies infection in the Mexican free-tail bat. J. infect. Dis. *117:* 82–90 (1967).

BAER, G. M. and WOODALL, D. F.: Bat salivary gland virus carrier state in a naturally infected Mexican free-tail bat. Amer. J. trop. Med. Hyg. *15:* 769–771 (1966).

BALDUCCI, M.: Personal commun. (1968).

BARBOUR, R. W. and DAVIS, W. H.: Bats of America (University Press of Kentucky, Lexington 1969).

BARKER, W. H. and FRANCIS, B. J.: Rabies in the Northwest. Epidemiology 1930–1969. Northwest Med. *71:* 179–183 (1972).

BEAUREGARD, M.: Bat rabies in Canada 1963–1967. Canad. J. comp. Med. *33:* 220–226 (1969).

BELL, J. F.; HADLOW, W. J., and JELLISON, W. L.: Chiropteran rabies in Montana. Publ. Hlth Rep., Wash. *70:* 991–994 (1955).

BELL, J. F.; LODMELL, D. L.; MOORE, G. J., and RAYMOND, G. H.: Rabies virus isolation from a bat in Montana in midwinter. Publ. Hlth Rep., Wash. *81:* 761–762 (1966).

BELL, J. F. and MOORE, G. J.: Rabies virus isolated from brown fat of naturally infected bats. Proc. Soc. exp. Biol. Med. *103:* 140–142 (1960).

BELL, J. F.; MOORE, G. J., and RAYMOND, G. H.: Protracted survival of a rabies-infected insectivorous bat after infective bite. Amer. J. trop. Med. Hyg. *18:* 61–66 (1969).

BELL, J. F. and THOMAS, L. A.: A new virus, 'MML', enzootic in bats *(Myotis lucifugus)* of Montana. Amer. J. trop. Med. Hyg. *13:* 607–612 (1964).

BELLAMY, R. E.; REEVES, W. C., and SCRIVANI, R. P.: Relationships of mosquito vectors to winter survival of encephalitis viruses. II. Under experimental conditions. Amer. J. Hyg. *67:* 90–100 (1958).

BELLAMY, R. E.; REEVES, W. C., and SCRIVANI, R. P.: Experimental cyclic transmission of Western equine encephalitis virus in chickens and *Culex tarsulis* through a year. Amer. J. Epidem. *85:* 282–296 (1967).

BLACKMORE, J. S. and WINN, J. F.: A winter isolation of western equine encephalitis virus from hibernating *Culex tarsalis* Coquillett. Proc. Soc. exp. Biol. Med. *91:* 146–148 (1956).

BLATTNER, R. J. and HEYS, F.: Role of viruses in the etiology of congenital malformations. Progr. med. Virol. *3:* 311–362 (1961).

BOEHME, D. H.: The reticuloendothelial system and non-specific resistance. Ann. N. Y. Acad. Sci. *88:* 172–183 (1960).

BOULGER, L. R. and PORTERFIELD, J. S.: Isolation of a virus from Nigerian fruit bats. Trans. roy. Soc. trop. Med. Hyg. *52:* 421–424 (1958).

BRÉS, P. et CHAMBON, L.: Isolement à Dakar d'une souche d'arbovirus à partir des glandes salivaires de chauve-souris. Ann. Inst. Pasteur *104:* 705–711 (1963).

BRÉS, P. et CHAMBON, L.: Techniques pour l'étude de l'infestation naturelle des chauves-souris par les arbovirus. Ann. Inst. Pasteur *107:* 34–43 (1964).

BRÉS, P.; WILLIAMS, M. C.; SIMPSON, D. I. H., and SANTOS, D. F.: Virology. Laboratory studies. Identification of viruses isolated in Senegal. East Africa Virus Res. Inst. Rep. 1963. *13:* 22–24 (1964).

BRYGOO, E. R. et DODIN, A.: Action de la cortisone sur l'évolution de la rage chez la souris. Ann. Inst. Pasteur *92:* 282–285 (1957).

BUESCHER, E. L.; SCHERER, W. F.; GROSSBERG, S. E.; CHANOCK, R., and VAN BUREN, P.: Immunologic studies of Japanese encephalitis virus in Japan. I. Antibody responses following overt infection of man. J. Immunol. *83:* 582–593 (1959a).

BUESCHER, E.; SCHERER, W.; ROSENBERG, M., and McCLURE, H. E.: Immunologic studies of Japanese encephalitis virus in Japan. III. Infection and antibody responses of birds. J. Immunol. *83:* 605–613 (1959b).

BUGBEE, L. M.; LIKE, A. A., and STEWART, R. B.: The effects of cortisone on intradermally induced vaccinia infection in rabbits. J. infect. Dis. *106:* 166–173 (1960).

BURNS, K. F.: Congenital Japanese B encephalitis infection of swine. Proc. Soc. exp. Biol. Med. *75:* 621–625 (1950).

BURNS, K. F.; FARINACCI, C. J.; MURNANE, T. G., and SHELTON, D. F.: Insectivorous bats naturally infected with rabies in the southwestern United States. Amer. J. publ. Hlth 46: 1089–1097 (1956).

BURNS, K. F.; FARINACCI, C. J., and SHELTON, D. F.: Virus of bats antigenically related to group B arthropod-borne encephalitis viruses. Amer. J. clin. Path. 27: 257–264 (1957).

BURNS, K. F.; SHELTON, D. F., and GROGAN, E. W.: Bat rabies. Experimental host transmission studies. Ann. N. Y. Acad. Sci. 70: 452–466 (1958).

BURTON, A. N.; McLINTOCK, J., and REMPEL, J. G.: Western equine encephalitis virus in Saskatchewan garter snakes. Science 154: 1029–1031 (1966).

CALISHER, C. H.; MANESS, K. S. C.; LORD, R. D., and COLEMAN, P. H.: Identification of two South American strains of Eastern equine encephalomyelitis virus from migrant birds captured on the Mississippi Delta. Amer. J. Epidem. 94: 172–178 (1971).

CAREY, D. E.; REUBEN, R.; MYERS, R. M., and GEORGE, S.: Japanese encephalitis studies in Vellore, South India. IV. Search for virological and serological evidence of infection in animals other than man. Indian J. med. Res. 56: 1340–1352 (1968).

CARINI, A.: Sur une grande epizootie de rage. Ann. Inst. Pasteur 25: 843–846 (1911).

CHAMBERLAIN, R. W.: Arboviruses, the arthropod-borne animal viruses. Curr. Topics Microbiol. Immunol. 42: 38–58 (1968).

CHAMBERLAIN, R. W.; KISSLING, R. E.; STAMM, D. D., and SUDIA, W. D.: Virus of St. Louis encephalitis in three species of wild birds. Amer. J. Hyg. 65: 110–118 (1957).

CHAMBERLAIN, R. W. and SUDIA, W. D.: Mechanisms of transmission of viruses by mosquitoes. Annu. Rev. Entomol. 6: 371–390 (1961).

CHANG, I.-C.: Studies on Japanese B encephalitis in cold-blooded animals. Pediatrics 4: 27–49 (1958).

CHIPPAUX, A. et CHIPPAUX-HYPPOLITE, C.: Une souche d'arbovirus isolée à Bangui à partir de glandes salivaires de chauves-souris. Bull. Soc. Path. exot. 58: 164–169 (1965).

CONSTANTINE, D. G.: Great bat colonies attract predators. Bull. nat. speleol. Soc. 10: 100 (1948).

CONSTANTINE, D. G.: A program for maintaining the free-tail bat in captivity. J. Mamm. 33: 395–397 (1952).

CONSTANTINE, D. G.: Transmission experiments with bat rabies isolates. Reaction of certain Carnivora, opossum, and bats to intramuscular inoculations of rabies virus isolated from free-tailed bats. Amer. J. vet. Res. 27: 16–19 (1966a).

CONSTANTINE, D. G.: Transmission experiments with bat rabies isolates. Bite transmission of rabies to foxes and coyote by free-tailed bats. Amer. J. vet. Res. 27: 20–23 (1966b).

CONSTANTINE, D. G.: Rabies transmission by air in bat caves. U.S. Publ. Hlth Serv. Publ. No. 1617 (U.S. Public Health Service, Washington 1967a).

CONSTANTINE, D. G.: Bat rabies in the southwestern United States. Publ. Hlth Rep., Wash. 82: 867–888 (1967b).

CONSTANTINE, D. G.: Bats in relation to the health, welfare and economy of man; in WIMSATT Biology of bats, vol. II, pp. 319–449 (Academic Press, New York 1970).

CONSTANTINE, D. G.; SOLOMON, G. C., and WOODALL, D. F.: Transmission experiments with bat rabies isolates. Responses of certain carnivores and rodents to rabies viruses from four species of bats. Amer. J. vet. Res. 29: 181–190 (1968).

CONSTANTINE, D. G. and WOODALL, D. F.: Latent infection of Rio Bravo virus in salivary glands of bats. Publ. Hlth Rep., Wash. *79:* 1033–1039 (1964).

CONSTANTINE, D. G. and WOODALL, D. F.: Transmission experiments with bat rabies isolates. Reactions of Carnivora, opossum, rodents, and bats to rabies virus of red bat origin when exposed by bat bite or by intramuscular inoculation. Amer. J. vet. Res. *27:* 24–32 (1966).

CORREA-GIRON, E. P.; ALLEN, R., and SULKIN, S. E.: The infectivity and pathogenesis of rabiesvirus administered orally. Amer. J. Epidem. *91:* 203–215 (1970).

CORREA-GIRON, P.; CALISHER, C. H., and BAER, G. M.: Epidemic strain of Venezuelan equine encephalomyelitis virus from a vampire bat captured in Oaxaca, Mexico, 1970. Science *175:* 546–547 (1972).

CORRISTAN, E. C.; LAMOTTE, L. C., jr., and SMITH, D. G.: Susceptibility of bats to certain encephalitis viruses. Fed. Proc. *15:* 584 (1956).

CRAIG, J. M. and SSENKUBUGE, Y.: Arbovirus isolation studies. Isolations from bats, Uganda collections. East Africa Virus Res. Inst. Rep. 1967. *17:* 20–21 (1968).

CROSS, J. H.; LIEN, J. C.; HUANG, W. C.; LIEN, S. C.; CHIU, S. F.; KUO, J.; CHU, H. H., and CHANG, Y. C.: Japanese encephalitis virus surveillance in Taiwan. II. Isolations from mosquitoes and bats in Taipei area 1969–1970. J. Formosan med. Ass. *70:* 681–686 (1971).

DALLDORF, G.: The Coxsackie viruses. Bull. N.Y. Acad. Med. *26:* 329–335 (1950).

DALQUEST, W. W. and WALTON, D. W.: Diurnal retreats of bats; in SLAUGHTER and WALTON About bats. A chiropteran symposium, pp. 162–187 (Southern Methodist University Press, Dallas 1970).

DANIELS, J. B.; STUART, G.; WHEELER, R. E.; GIFFORD, C.; AHEARN, J. P.; PHILBROOK, F. R.; HAYES, R. O., and MACCREADY, R. A.: A search for encephalitis and rabies in bats of eastern Massachusetts. New Engl. J. Med. *263:* 516–520 (1960).

DAVIS, R. B.; HERREID, C. F., II, and SHORT, H. L.: Mexican free-tailed bats in Texas. Ecological Monogr. *32:* 311–346 (1962).

DAVIS, W. H.: Hibernation. Ecology and physiological ecology; in WIMSATT Biology of bats, vol. I, pp. 265–300 (Academic Press, New York 1970).

DEMPSTER, G.; GRODUMS, E. I., and SPENCER, W. A.: Experimental Coxsackie B-3 infection in the hibernating squirrel and bat. Canad. J. Microbiol. *7:* 587–594 (1961).

DIJKGRAAF, S.: Spallanzani's unpublished experiments on the sensory basis of object perception in bats. Isis *51:* 9–20 (1960).

DONALDSON, A. I.: Bats as possible maintenance hosts for vesicular stomatitis virus. Amer. J. Epidem. *92:* 132–136 (1970).

DOWNS, W. G.; ANDERSON, C. R.; SPENCE, L.; AITKEN, T. H. G., and GREENHALL, A. H.: Tacaribe virus, a new agent isolated from *Artibeus* bats and mosquitoes in Trinidad, West Indies. Amer. J. trop. Med. Hyg. *12:* 640–646 (1963).

EBERT, J. D. and WILT, F. H.: Animal viruses and embryos. Quart. Rev. Biol. *35:* 261–312 (1960).

EHRENKRANZ, N. J.; POND, W. I., and DANAUSKAS, J. X.: Recurring St. Louis encephalitis virus infection at an urban homesite. New Engl. J. Med. *277:* 12–16 (1967).

EISENTRAUT, M.: Berichte und Ergebnisse von Markierungsversuchen und Fledermäusen in Deutschland und Österreich. Bonn. zool. Beitr. *11:* 7–263 (1960).

EL-SABBAN, M. S.; FOUAD, M. S.; HUSSEIN, N., and SADEK, B.: Bats and their relation to rabies in the United Arab Republic. Bull. Office int. Epizoot. *67:* 543–557 (1967).

EMMONS, R. W. and LENNETTE, E. H.: Isolation of St. Louis encephalitis virus from a naturally-infected gray fox *Urocyon cinereoargenteus.* Proc. Soc. exp. Biol. Med. *125:* 443–447 (1967).

ENRIGHT, J. B.: Bats, and their relation to rabies. Annu. Rev. Microbiol. *10:* 369–392 (1956).

ENRIGHT, J. B.; SADLER, W. W.; MOULTON, J. E., and CONSTANTINE, D.: Isolation of rabies virus from an insectivorous bat *(Tadarida mexicana)* in California. Proc. Soc. exp. Biol. Med. *89:* 94–96 (1955).

FERNANDES, M. V.; WIKTOR, T. J., and KOPROWSKI, H.: Mechanisms of the cytopathic effect of rabies virus in tissue culture. Virology *21:* 128–131 (1963).

FISCHMAN, H. R. and WARD, F. E., III: Oral transmission of rabies virus in experimental animals. Amer. J. Epidem. *88:* 132–138 (1968).

FUKUMI, H.; HAYASHI, K.; MIFUNE, K.; UJIIYE, A.; SUENAGA, O.; FUTATSUKI, K.; MATSUO, S., and MIYAGI, I.: Epidemiological studies on viral and parasitic disease and vector insects in East Africa. I. Virological studies especially on the epidemiology of arboviruses. Trop. Med., Jap. *9:* 127–135 (1967).

GATES, W. H.: Keeping bats in captivity. J. Mamm. *17:* 268–273 (1936).

GATES, W. H.: Raising the young of red bats on artificial diet. J. Mamm. *19:* 461–464 (1938).

GEBHARDT, L. P. and STANTON, G. J.: The role of poikilothermic hosts as virus reservoirs. Proc. Symp. held at the 11th Pacific Science Congr., Tokyo 1966, vol. 20, pp. 30–34.

GEBHARDT, L. P.; ST. JEOR, S. C.; STANTON, G. J., and STRINGFELLOW, D. A.: Ecology of western encephalitis virus. Proc. Soc. exp. Biol. Med. *142:* 731–733 (1973).

GILTNER, L. T. and SHAHAN, M. S.: Transmission of infectious equine encephalomyelitis in mammals and birds. Science *78:* 63–64 (1933).

GIRARD, K. F.; HITCHCOCK, H. B.; EDSALL, G., and MACCREADY, R. A.: Rabies in bats in southern New England. New Engl. J. Med. *272:* 75–80 (1965).

GLASS, B. P.: Feeding mechanisms of bats; in SLAUGHTER and WALTON About bats. A chiropteran symposium, pp. 84–92 (Southern Methodist University Press, Dallas 1970).

GLEDHILL, A. W.: Sparing effect of serum from mice treated with endotoxin upon certain murine virus diseases. Nature, Lond. *183:* 185–186 (1959).

GOIDL, E. A.; PAUL, W. E.; SISKIND, G. W., and BENACERRAF, B.: The effect of antigen dose and time after immunization on the amount and affinity of anti-hapten antibody. J. Immunol. *100:* 371–375 (1968).

GOLDFIELD, M.: Personal commun. (1968).

GOLDWASSER, R. A.; KISSLING, R. E.; CARSKI, T. R., and HOSTY, T. S.: Fluorescent antibody staining of rabies virus antigens in the salivary glands of rabid animals. Bull. Wld Hlth Org. *20:* 579–588 (1959).

GOOD, R. A. and MOLLER, G.: In LANDY and BRAUN Immunological tolerance. A reassessment of mechanisms of immune response, p. 215 (Academic Press, New York 1969).

GOODMAN, G. T. and KOPROWSKI, H.: Study of the mechanism of innate resistance to virus infection. J. cell. comp. Physiol. *59:* 333–373 (1962).

GRAYSON, M. A. and GALINDO, P.: Epidemiologic studies of Venezuelan equine encephalitis virus in Almirante, Panama. Amer. J. Epidem. *88:* 80–96 (1968).

GREENHALL, A. M.: Lucha contra los murcielagos vampiros. Estudio y proyecto de programa para America Latina. Bol. Off. Sanit. Panamer. *71:* 231–247 (1971).

GRIFFIN, D. R.: Listening in the dark (Yale Univ. Press, New Haven 1958).

GRIFFIN, D. R.: Migrations and homing of bats; in WIMSATT Biology of bats, vol. I, pp. 233–264 (Academic Press, New York 1970).

GRIFFIN, R. and HITCHCOCK, B.: Probable 24-year longevity records for *Myotis lucifugus.* J. Mamm. *46:* 332 (1965).

HAMMON, W. McD.: The viral encephalitides. Ann. N. Y. Acad. Sci. *70:* 292–293 (1958).

HANSON, R. P.: The natural history of vesicular stomatitis. Bact. Rev. *16:* 179–204 (1952).

HARTRIDGE, H.: The avoidance of objects by bats in their flight. J. Physiol., Lond. *54:* 54–57 (1920).

HATTEN, B. A.; ALLEN, R., and SULKIN, S. E.: Immune response in Chiroptera to bacteriophage ⌀X174. J. Immunol. *101:* 141–150 (1968).

HATTEN, B. A.; ALLEN, R., and SULKIN, S. E.: Studies on the immune capabilities of Chiroptera. I. Quantitative and qualitative nature of the immune responses in bats to bacteriophage ⌀X174. J. Immunol. *105:* 872–878 (1970).

HATTWICK, M. A. W.; WEIS, T. T.; STECHSCHULTE, J.; BAER, G. M., and GREGG, M. B.: Recovery from rabies: a case report. Ann. intern. Med. *76:* 931–942 (1972).

HAUPT, H. und REHAAG, H.: Durch Fledermäuse verbreitete seuchenhafte Tollwut unter Viehbeständen in Santa Catharina (Süd-Brasilien). Z. Infektionskr. *22:* 104–127 (1921).

HAVLIK, O. and KOLMAN, J. M.: Detection of antibodies against tickborne encephalitis in certain domestic bats. Čs. Mikrobiol. *6:* 241–244 (1957).

HAYWARD, J. S. and LYMAN, C. P.: Nonshivering heat production during arousal from hibernation and evidence for the contribution of brown fat; in FISHER Mammalian hibernation, vol. III, pp. 346–355 (Oliver & Boyd, Edinburgh 1967).

HENDERSON, B. E.: TUKEI, P. M.; LULE, M., and MUTERE, F. A.: Arbovirus identification studies: isolations from bats. East Africa Virus Res. Inst. Rep. 1967. *17:* 27–28 (1968).

HENSHAW, R. E.: Thermoregulation in bats; in SLAUGHTER and WALTON About bats. A chiropteran symposium, pp. 188–232 (Southern Methodist University Press, Dallas 1970).

HOCK, R. J.: The metabolic rates and body temperatures of bats. Biol. Bull. *101:* 289–299 (1951).

HOLLINGER, F. B. and PAVRI, K. M.: Bat parainfluenza virus. Immunological, chemical, and physical properties. Amer. J. trop. Med. Hyg. *20:* 131–138 (1971).

HUGHES, T. P. and PERLOWAGORA, A.: The reaction of certain species of bats to yellow fever virus. Amer. J. trop. Med. Hyg. *28:* 101–105 (1948).

HULL, T. G.: Diseases transmitted from animals to man, 4th ed. (Thomas, Springfield 1955).

HUMPHREY, C. L.; KEMP, G. E., and WOOD, E. G.: A fatal case of rabies in a woman bitten by an insectivorous bat. Publ. Hlth Rep., Wash. *75:* 317–326 (1960).

HURLBUT, H. S.: The transmission of Japanese B encephalitis by mosquitoes after experimental hibernation. Amer. J. Hyg. *51:* 265–268 (1950).

IMAM, Z. E. and HAMMON, W. McD.: Susceptibility of hamsters to peripherally inoculated Japanese B and St. Louis viruses following cortisone, X-ray, trauma. Proc. Soc. exp. Biol. Med. *95:* 6–11 (1957).

IRONS, J. V.; EADS, R. B.; GRIMES, J. E., and CONKLIN, A.: The public health importance of bats. Texas Rep. Biol. Med. *15:* 292–298 (1957).

IRONS, J. V.; EADS, R. B.; SULLIVAN, T., and GRIMES, J. E.: Current status of rabies in Texas. Texas Rep. Biol. Med. *12:* 489–499 (1954).

ITO, T. and SAITO, S.: Susceptibility of bats to Japanese B encephalitis virus. Jap. J. Bact. *7:* 617–622 (1952).

JAROSLOW, B. N.: Development of the secondary hemolysin response in hibernating ground squirrels *(Citellus tridecemlineatus)*. Proc. nat. Acad. Sci., Wash. *61:* 69–76 (1968).

JELESIC, Z. und JELESIC, R.: Histopathologische Eigenschaften des Tollwutvirus der Fledermäuse in Jugoslawien. Arch. Hyg. Bakt. *143:* 619–623 (1959).

JEPSEN, G. L.: Early Eocene bat from Wyoming. Science *154:* 1333–1339 (1966).

JEPSEN, G. L.: Bat origins and evolution; in WIMSATT Biology of bats, vol. I, pp. 1–64 (Academic Press, New York 1970).

JOHANSSON, B.: Brown fat. A review. Metab. clin. Exp. *8:* 221–240 (1959).

JOHNSON, H. N.: The Rio Bravo virus. Virus identified with group B arthropod-borne viruses by hemagglutination inhibition and complement fixation tests. Proc. 9th Pacific Science Congr., 1957. *17:* 39 (1962).

JOHNSON, H. N.: Diseases derived from wildlife. Calif. Hlth *23:* 35–39 (1965).

JOHNSON, H. N.: Ecological implications of antigenically related mammalian viruses for which arthropod vectors are unknown and avian associated soft tick viruses. Jap. J. med. Sci. Biol. *20:* suppl., pp. 160–166 (1967).

KAPLAN, M. M.: Epidemiology of rabies. Nature, Lond. *221:* 421–425 (1969).

KARABATSOS, N.: Characterization of viruses isolated from bats. Amer. J. trop. Med. Hyg. *18:* 803–810 (1969).

KARSTAD, L.: Reptiles as possible reservoir hosts for eastern encephalitis virus. Trans. 26th N. Amer. Wildlife Conf., 1961, pp. 186–202.

KARSTAD, L. H. and HANSON, R. P.: Infections in wildlife with the viruses of vesicular stomatitis and eastern equine encephalomyelitis. Trans. 23rd N. Amer. Wildlife Conf., pp. 175–186, 1958.

KASAHARA, S.; UEDA, M.; OKAMOTO, Y.; YOSHIDA, S.; HAMANO, R., and YAMADA, R.: Experimental studies on epidemic encephalitis. Transmission test of Japanese encephalitis in 1935 and some characteristics of infectious agent. Kitasato Arch. exp. Med. *13:* 48–65 (1936).

KENT, J. R. and FINEGOLD, S. M.: Human rabies transmitted by the bite of a bat. With comments on the duck-embryo vaccine. New Engl. J. Med. *263:* 1058–1065 (1960).

KILBOURNE, E. D. and HORSFALL, F. L.: Lethal infection with Coxsackie virus of adult mice given cortisone. Proc. Soc. exp. Biol. Med. *77:* 135–138 (1951).

KISSLING, R. E.: Growth of rabies virus in non-nervous tissue culture. Proc. Soc. exp. Biol. Med. *98:* 223–225 (1958).

KISSLING, R. E.: The arthropod-borne viruses of man and other animals. Annu. Rev. Microbiol. *14:* 261–282 (1960).

KISSLING, R. E. and REESE, D. R.: Anti-rabies vaccine of tissue culture origin. J. Immunol. *91:* 362–368 (1963).

KISSLING, R. E.; STAMM, D. D.; CHAMBERLAIN, R. W., and SUDIA, W. D.: Birds as winter
hosts for eastern and western equine encephalomyelitis viruses. Amer. J. Hyg. *66:*
42–47 (1957).

KOLMAN, J. M.: FISCHER, J., and HAVLIK, O.: Experimental infection of bat species
Myotis myotis Borkhausen with the Czechoslovak tick-borne encephalitis virus. Acta
Univ. Carol. med. *2:* 147–180 (1960).

KOOPMAN, K. F.: Zoogeography of bats; in SLAUGHTER and WALTON About bats. A
chiropteran symposium, pp. 29–50 (Southern Methodist University Press, Dallas 1970).

KOOPMAN, K. F. and JONES, J. K.: Classification of bats; in SLAUGHTER and WALTON
About bats. A chiropteran symposium, pp. 22–28 (Southern Methodist University
Press, Dallas 1970).

KRUTZSCH, P. H. and SULKIN, S. E.: The laboratory care of the Mexican free-tailed bat.
J. Mamm. *39:* 262 (1958).

KUMM, H. W.: Yellow fever transmission experiments with South American bats. Ann.
trop. Med. Parasit. *26:* 207–213 (1932).

LaCHANCE, J. P. and PAGE, E.: Hormonal factors influencing fat deposition in the interscapular brown adipose tissue of the white rat. Endocrinology *52:* 57–64 (1953).

LAEMMERT, H. W., jr.; DE CASTRO FERREIRA, L., and TAYLOR, R. M.: An epidemiological
study of jungle yellow fever in an endemic area of Brazil. II. Investigation of vertebrate
hosts and arthropod vectors. Amer. J. trop. Med. Hyg. *26:* 23–69 (1946).

LaMOTTE, L. C., jr.: Japanese B encephalitis in bats during simulated hibernation. Amer.
J. Hyg. *67:* 101–108 (1958).

LENNETTE, E. H.: General principles underlying laboratory diagnosis of viral and rickettsial infections; in LENNETTE and SCHMIDT Diagnostic procedures for viral and rickettsial
infections, 4th ed., pp. 1–65 (Amer. Publ. Hlth Ass., New York 1969).

LENNETTE, E. H.; SOAVE, O. A.; NAKAMURA, K., and KELLOGG, G. H., jr.: Fatal human
case of rabies following bite of rabid bat *(Lasionycteris noctivagans)*. Isolation and
identification of virus from vector and victim. J. Lab. clin. Med. *55:* 89–93 (1960).

LEONARD, L. L.; ALLEN, R., and SULKIN, S. E.: Studies with cultured brown adipose
tissue. III. A cytochemical study of rabies virus inclusion bodies. J. infect. Dis. *117:*
121–128 (1967).

LEONARD, L. L.; ALLEN, R., and SULKIN, S. E.: Bat immunoglobulins formed in response
to experimental Japanese B encephalitis (JBE) virus infection. J. Immunol. *101:*
1168–1175 (1968).

LINDBERG, O.: Brown adipose tissue (American Elsevier, New York 1970).

LINHART, S. B.; CRESPO, R. F., and MITCHELL, G. C.: Control of vampire bats by topical
application of an anticoagulant, Chlorophacinone. Bol. Off. Sanit. Panamer. *6:* 31–38
(1972).

LORD, R. D.: Some vertebrate host aspects of arbovirus ecology. J. Wildlife Dis. *6:*
236–238 (1970).

LORD, R. D. and CALISHER, C. H.: Further evidence of southward transport of arboviruses by migratory birds. Amer. J. Epidem. *92:* 73–78 (1971).

LUBY, J. P.; SULKIN, S. E., and SANFORD, J. P.: The epidemiology of St. Louis encephalitis.
A review. Annu. Rev. Med. *20:* 329–350 (1969).

LUMSDEN, W. H. R.; WILLIAMS, M. C., and MASON, P. J.: A virus from insectivorous bats
in Uganda. Ann. trop. Med. Parasit. *55:* 389–397 (1961).

LYMAN, C. P.: Thermoregulation and metabolism in bats; in WIMSATT Biology of bats, vol. I, pp. 301–330 (Academic Press, New York 1970).

MAIN, A. J., jr.: Natural and experimental infections of eastern equine encephalomyelitis and other arboviruses in colonial bats (Chiroptera: Vespertilionidae of New England); M.S. thesis, Univ. of Massachusetts (1970).

MALAGA ALBA, A.: Vampire bat as a carrier of rabies. Amer. J. publ. Hlth 44: 909–918 (1954).

MATUMOTO, M.: Mechanism of perpetuation of animal viruses in nature. Bact. Rev. 33: 404–418 (1969).

MCKENNA, J. M. and MUSACCHIA, X. J.: Antibody formation in hibernating ground squirrels (Citellus tridecemlineatus). Proc. Soc. exp. Biol. Med. 129: 720–724 (1968).

MEDOVY, H.: Western equine encephalomyelitis in infants. J. Pediat. 22: 308–324 (1943).

MENAKER, M.: Hibernation-hypothermia. An annual cycle of response to low temperature in the bat Myotis lucifugus. J. cell. comp. Physiol. 59: 163–173 (1962).

METSELAAR, D.; WILLIAMS, M. C.; SIMPSON, D. I. H.; WEST, R., and MUTURE, F. A.: Mount Elgon bat virus. A hitherto undescribed virus from Rhinolophus hildebrandtii eloquens K. Anderson. Arch ges. Virusforsch. 26: 183–193 (1969).

MIDDLEBROOKS, B. L.: The course of experimental arbovirus infection in Chiroptera; Ph.D. Diss., Univ. of Texas Southwestern Medical School, Dallas 1966.

MIDDLEBROOKS, B. L.; ALLEN, R., and SULKIN, S. E.: The effects of 6-thioguanine and endotoxin on experimental St. Louis and Japanese B encephalitis virus infections in Chiroptera. Proc. Soc. exp. Biol. Med. 127: 886–890 (1968).

MIDDLEBROOKS, B.; ALLEN, R., and SULKIN, S. E.: Studies of arthropod-borne virus infections in Chiroptera. V. Characteristics of lines of Japanese B encephalitis virus developed by serial passage in big brown bats (Eptesicus f. fuscus) maintained at different environmental temperatures. Amer. J. trop. Med. Hyg. 18: 115–122 (1969).

MILLER, G. S.: Families and genera of bats. U.S. Nat. Mus. Bull. 57: 1–282 (1907).

MIURA, T. and KITAOKA, M.: Immunological epidemiology of Japanese encephalitis in Hokkaido. Virus 5: 62–73 (1955).

MIURA, T. and SCHERER, W. F.: Comparison of chicken embryonic cell cultures and mice for detecting neutralizing antibody to Japanese encephalitis virus. Use of microcultures for virus- and serum-dilution neutralization tests. Amer. J. Hyg. 76: 197–208 (1962).

MIURA, T.; TOYOKAWA, K.; ALLEN, R., and SULKIN, S. E.: Studies of arthropod-borne virus infections in Chiroptera. VII. Serologic evidence of natural Japanese B encephalitis virus infection in bats. Amer. J. trop. Med. Hyg. 19: 88–93 (1970).

MOHOS, S. C.: Bats as laboratory animals. Anat. Rec. 139: 369–378 (1961).

MORRISON, P.: Body temperatures in some Australian mammals. I. Chiroptera. Biol. Bull. 116: 484–497 (1959).

MURPHY, F. A. and FIELDS, B. N.: Kern Canyon virus. Electron microscopic and immunological studies. Virology 33: 625–637 (1967).

MURPHY, F. A.; SHOPE, R. E.; METSELAAR, D., and SIMPSON, D. I. H.: Characterization of Mount Elgon bat virus, a new member of the rhabdovirus group. Virology 40: 288–297 (1970).

NIKOLIC, M. and JELESIC, Z.: Isolation of rabies virus from insectivorous bats in Yugoslavia. Bull. Wld Hlth Org. 14: 801–804 (1956).

NIKOLITSCH, V. M.: Das Virus der Fledermäuse und die Endephalomyelitis des Menschen. Arch. Hyg. Bakt. *143:* 607–618 (1959).

NOSEK, J.; GRESIKOVA, M., and REHACEK, J.: Persistence of tick-borne encephalitis virus in hibernating bats. Acta virol. *5:* 112–116 (1961).

OYA, A.; OKUNO, T.; OGATA, T., and KOBAYASHI, I.: Akabane, a new arbor virus isolated in Japan. Jap. J. med. Sci. Biol. *14:* 101–108 (1961).

PAPPENHEIMER, A. M.; DANIELS, J. B.; CHEEVER, F. S., and WELLER, T. H.: Lesions caused in suckling mice by certain viruses isolated from cases of so-called non-paralytic poliomyelitis and of pleurodynia. J. exp. Med. *92:* 169–190 (1950).

PAUL, S. D.; RAJAGOPALAN, P. K., and SREENIVASAN, M. A.: Isolation of the West Nile virus from the frugivorous bat, *Rousettus leschenaulti*. Indian J. med. Res. *58:* 1169–1171 (1970).

PAVRI, K. M. and SINGH, K. R. P.: Demonstration of antibodies against the virus of Kyasanur Forest disease (KFD) in the frugivorous bat *Rousettus leschenaulti,* near Poona, India. Indian J. med. Res. *53:* 956–961 (1965).

PAVRI, K. M. and SINGH, K. R. P.: Kyasanur Forest disease virus infection in the frugivorous bat, *Cynopterus sphinx*. Indian J. med. Res. *56:* 1202–1204 (1968).

PAVRI, K. M.; SINGH, K. R. P., and HOLLINGER, F. B.: Isolation of a new parainfluenza virus from a frugivorous bat, *Rousettus leschenaulti,* collected at Poona, India. Amer. J. trop. Med. Hyg. *20:* 125–130 (1971).

PAWAN, J. L.: The transmission of paralytic rabies in Trinidad by the vampire bat *(Desmodus rotundus murinus* Wagner, 1840). Ann. trop. Med. Parasit. *30:* 101–130 (1936a).

PAWAN, J. L.: Rabies in the vampire bat of Trinidad, with special reference to the clinical course and latency of infection. Ann. trop. Med. Parasit. *30:* 401–422 (1936b).

PETERSON, R.: Silently, by night (McGraw-Hill, New York 1964).

PHILLIPS, C. A. and MELNICK, J. L.: Urban epidemic encephalitis in Houston caused by a group B arbovirus (SLE). Progr. med. Virol., vol. 9, pp. 159–175 (Karger, Basel 1967).

PIKE, R. M.: Antibody heterogeneity and serological reactions. Bact. Rev. *31:* 157–174 (1967).

PITZSCHKE, H.: Tollwut bei einer Breitflügel-Fledermaus *(Eptesicus serotinus)* in Thüringen. Zbl. Bakt. *196:* 411–415 (1965).

PRIAS-LANDINEZ, E.: Personal commun. (1966).

QUEIROZ LIMA, E.: A transmissao da raiva dos herbivoros pelos morcegos hematophagos da familia Desmodontidae. Rev. Dept. nac. Prod. Animal *1:* 165 (1934).

RAJAGOPALAN, P. K.; PAUL, S. D., and SREENIVASAN, M. A.: Isolation of Kyasanur Forest disease virus from the insectivorous bat, *Rhinolophus rouxi* and from *Ornithodoros* ticks. Indian J. med. Res. *57:* 805–808 (1969).

RAMAGE, M. C.: Notes on keeping bats in captivity. J. Mamm. *28:* 60–62 (1947).

RAO, T. R.: Annual report (Virus Research Centre, Poona 1968).

RAUSCH, R.: Some observations on rabies in Alaska, with special reference to wild Canidae. J. Wildlife Mgmt. *22:* 246–260 (1958).

REAGAN, R. L. and BRUECKNER, A. L.: Transmission of a strain of rabiesvirus to the large brown bat *(Eptesicus fuscus)* and to the cave bat *(Myotis lucifugus)*. Cornell Vet. *41:* 295–298 (1951).

REAGAN, R. L.; DELAHA, E. C., and BRUECKNER, A. L.: Response of the cave bat to several strains of rabies virus. Cornell Vet. *44:* 318–321 (1954).

REAGAN, R. L.; YANCEY, F. S., and BRUECKNER, A. L.: Transmission of rabies from artificially infected bats to Syrian hamsters. Amer. med. Ass. Arch. Path. *63:* 278–280 (1957).

REAGAN, R. L.; DELAHA, E. C.; COOK, S. R., and BRUECKNER, A. L.: Response of the cave bat *(Myotis lucifugus)* to the Lansing strain of poliomyelitis virus. Cornell Vet. *44:* 449–452 (1954).

REED, W.; Yellow fever. A compilation of various publications. Document No. 822, Government Printing Office, Washington (1911).

REEVES, W. C.: Overwintering of arthropod-borne viruses. Progr. med. Virol., vol. 3, pp. 59–78 (Karger, Basel 1961).

REEVES, W. C.; BELLAMY, R. E., and SCRIVANI, R. P.: Relationships of mosquito vectors to winter survival of encephalitis viruses. I. Under natural conditions. Amer. J. Hyg. *67:* 78–89 (1958a).

REEVES, W. C.; HUTSON, G. A.; BELLAMY, R. E., and SCRIVANI, R. P.: Chronic latent infections of birds with Western equine encephalitis. Proc. Soc. exp. Biol. Med. *97:* 733–736 (1958b).

RODHAIN, J.: La recepticité de la rouséette epaulière *Epomophorus wahlbergi haldemani* (Hallowell) au virus amaril neurotrope. C. R. Soc. Biol. *123:* 1007–1009 (1936).

ROHITAYODHIN, S. and HAMMON, W. McD.: Studies on Japanese B encephalitis virus vaccines from tissue culture. II. Development of an attenuated strain of virus. J. Immunol. *89:* 589–597 (1962).

ROKUTANDA, H. K.: Relationship between viremia and interferon production of Japanese encephalitis virus. J. Immunol. *102:* 662–670 (1969).

ROWAN, L. C.: Recent work on dengue fever. Med. J. Austr. *2:* 530–533 (1957).

ROWAN, L. C. and O'CONNOR, J. L.: Relationship between some coastal fauna and arthropod-borne fevers of North Queensland. Nature, Lond. *179:* 786–787 (1957).

RUEGER, M. E.; OLSON, T. A., and PRICE, R. D.: Studies of potential avian, arthropod and mammalian hosts of mosquito-borne arboviruses in the Minnesota area. Amer. J. Epidem. *83:* 33–37 (1966).

RUSSELL, P. K.: Personal commun. (1968).

RUSSELL, P. K.; NISALKA, A.; SUKHAVACHANA, P., and VIVONA, S.: A plaque reduction test for dengue virus neutralizing antibodies. J. Immunol. *99:* 285–290 (1967).

SADLER, W. W. and ENRIGHT, J. B.: Effect of metabolic level of the host upon the pathogenesis of rabies in the bat. J. infect. Dis. *105:* 267–273 (1959).

SADLER, W. W. and TYLER, W. S.: Thyroidal activity in hibernating Chiroptera. I. Uptake of 131-I. Acta endocrin., Kbh. *34:* 586–596 (1960).

SALAUN, J. J.: Personal commun. (1970).

SCATTERDAY, J. E.: Bat rabies in Florida. J. amer. vet. med. Ass. *124:* 125 (1954).

SCHERER, W. F. and BUESCHER, E. L.: Ecologic studies of Japanese encephalitis virus in Japan. Amer. J. trop. Med. Hyg. *8:* 644–650 (1959).

SCHERER, W. F.; DICKERMAN, R. W.; LA FIANDRA, R. P.; CHIA, C. W., and TERRIAN, J.: Ecologic studies of Venezuelan encephalitis virus in southeastern Mexico. IV. Infections of wild mammals. Amer. J. trop. Med. Hyg. *20:* 980–988 (1971).

SCHERER, W. F.; IZUMI, T.; McCOWN, J., and HARDY, J. L.: Sagiyama virus. II. Some biologic, physical, chemical and immunologic properties. Amer. J. trop. Med. Hyg. *11:* 269–282 (1962).

Scherer, W. F.; Kitaoka, M.; Grossberg, S.; Okuno, T.; Ogata, T., and Chanock, R.: Immunologic studies of Japanese encephalitis virus in Japan. II. Antibody responses following inapparent human infection. J. Immunol. *83:* 594–604 (1959a).

Scherer, W.; Moyer, J., and Toshiaki, I.: Immunologic studies of Japanese encephalitis virus in Japan. V. Maternal antibodies, antibody responses and viremia following infection of swine. J. Immunol. *83:* 620–626 (1959b).

Serie, C.; Andral, L.; Poirier, A.; Lindrec, A. et Neri, P.: Etudes sur la fièvre jaune en Ethiopie. 6. Etude epidemiologique. Bull. Wld Hlth Org. *38:* 879–884 (1968).

Shah, K. V. and Daniel, R. W.: Attempts at experimental infection of the Indian fruit-bat *Pteropus giganteus* with chikungunya and dengue 2 viruses and antibody survey of bat sera for some viruses. Indian J. med. Res. *54:* 714–722 (1966).

Shepherd, R. C. and Williams, M. C.: Studies on viruses in East African bats (Chiroptera). 1. Hemagglutination inhibition and circulation of arboviruses. Zoonoses Res. *3:* 125–139 (1964).

Shimizu, T.; Kawakami, Y.; Fukuhara, S., and Matumoto, M.: Experimental still-birth in pregnant swine infected with Japanese encephalitis virus. Jap. J. exp. Med. *24:* 363–375 (1954).

Shinefield, H. R. and Townsend, T. E.: Transplacental transmission of Western equine encephalomyelitis. J. Pediat. *43:* 21–25 (1953).

Shiraki, H.: The neuropathology of encephalitis Japonica in humans especially from subchronic to chronic stage. Neuropath. Polska *4:* 419–448 (1966).

Shope, R. E.; Murphy, F. A.; Harrison, A. K.; Causey, O. R.; Kemp, G. E.; Simpson, D. I. H., and Moore, D. L.: Two African viruses serologically and morphologically related to rabies virus. J. Virol. *6:* 690–692 (1970).

Shwartzman, G.: The effect of ACTH and cortisone upon infection and resistance (Columbia University Press, New York 1953).

Sidman, R. L.: The direct effect of insulin on organ cultures of brown fat. Anat. Rec. *124:* 723–739 (1956).

Simpson, D. I. H. and O'Sullivan, J. P.: Studies on arboviruses and bats (Chiroptera) in East Africa. I. Experimental infection of bats and virus transmission attempts in *Aedes (Stegomyia) aegypti* (Linnaeus). Ann. trop. Med. Parasit. *62:* 422–431 (1968).

Simpson, D. I. H.; Williams, M. C.; O'Sullivan, J. P.; Cunningham, J. C., and Mutere, F. A.: Studies on arboviruses and bats (Chroptera) in East Africa. II. Isolation and haemagglutination-inhibition studies on bats collected in Kenya and throughout Uganda. Ann. trop. Med. Parasit. *62:* 432–440 (1968).

Sims, R. A.: A study of natural differences and induced alterations in rabiesvirus strains; Ph.D. Diss., Univ. of Texas Southwestern Medical School, Dallas (1969).

Sims, R.; Allen, R., and Sulkin, S. E.: Influence of pregnancy and cortisone treatment on brown adipose tissue in bats. Proc. Soc. exp. Biol. Med. *111:* 455–458 (1962).

Sims, R.; Allen, R., and Sulkin, S. E.: Studies on the pathogenesis of rabies in insectivorous bats. III. Influence of the gravid state. J. infect. Dis. *112:* 17–27 (1963).

Sims, R. A.; Allen, R., and Sulkin, S. E.: Influence of lipogenesis on growth of rabies virus in brown adipose tissue of experimentally infected hamsters. J. infect. Dis. *117:* 360–370 (1967).

SLAUGHTER, B. H.: Evolutionary trends of Chiropteran dentitions; in SLAUGHTER and WALTON About bats. A chiropteran symposium, pp. 51–83 (Southern Methodist University Press, Dallas 1970).

SLAUGHTER, B. H. and WALTON, D. W. (eds.): About bats (Southern Methodist University Press, Dallas 1970).

SMALLEY, R. L. and DRYER, R. L.: Brown fat. Thermogenic effect during arousal from hibernation in the bat. Science *140:* 1333–1334 (1963).

SMALLEY, R. L. and DRYER, R. L.: Brown fat in hibernation; in FISHER Mammalian hibernation, vol. III, pp. 325–345 (Oliver & Boyd, Edinburgh 1967).

SMITH, P. C.; LAWHASWASKI, K.; VICK, W. E., and STANTON, J. S.: Isolation of rabies virus from fruit bats in Thailand. Nature, Lond. *216:* 384 (1967).

SMITH, R. E.: Thermogenic activity of the hibernating gland in the cold-acclimated rat. Physiologist *4:* 113 (1961).

SMITH, R. E. and HORWITZ, B. A.: Brown fat and thermogenesis. Physiol. Rev. *49:* 330–425 (1969).

SOAVE, O. A.: Transmission of rabies to mice by ingestion of infected tissue. Amer. J. vet. Res. *27:* 44–46 (1966).

STAMM, D. D.; KISSLING, R. E., and EIDSEN, M. E.: Experimental rabies infection in insectivorous bats. J. infect. Dis. *98:* 10–14 (1956).

STAMM, D. D. and NEWMAN, R. J.: Evidence of southward transport of arboviruses from the U.S. by migratory birds. Anais Microbiol. *11:* 123–133 (1963).

STANLEY, N. F.; LEAK, P. J.; GRIEVE, G. M., and PARRET, D.: The ecology and epidemiology of reovirus. Austr. J. exp. Biol. med. Sci. *42:* 373–384 (1964).

STEWART, W. E., II: Mechanisms of persistence of Japanese encephalitis virus in bats and bat cell cultures; Ph.D. Diss., Univ. of Texas Southwestern Medical School, Dallas (1970).

STEWART, W. E., II; ALLEN, R., and SULKIN, S. E.: Persistent infection in bats and bat cell cultures with Japanese encephalitis virus. Bact. Proc., p. 193 (1969a).

STEWART, W. E., II; LUTSKUS, J. H., and SULKIN, S. E.: Chromosomal stability of bat cells in culture. Caryologia *23:* 163–170 (1970).

STEWART, W. E., II; SCOTT, W. D., and SULKIN, S. E.: Relative sensitivities of viruses to different species of interferon. J. Virol. *4:* 147–153 (1969b).

STONE, R. C.: Laboratory care of little brown bats at thermal neutrality. J. Mamm. *46:* 681–682 (1965).

STRODE, G. K.: Yellow fever (McGraw-Hill, New York 1951).

SULKIN, S. E.: The bat as a reservoir of viruses in nature. Progr. med. Virol., vol. 4, pp. 157–207 (Karger, Basel 1962).

SULKIN, S. E.: Chiropteran (bat) rabies in North America. Hawaii med. J. *25:* 149–153 (1965).

SULKIN, S. E.; ALLEN, R.; MIURA, T., and TOYOKAWA, K.: Studies of arthropod-borne virus infections in Chiroptera. VI. Isolation of Japanese B encephalitis virus from naturally infected bats. Amer. J. trop. Med. Hyg. *19:* 77–87 (1970).

SULKIN, S. E.; ALLEN, R., and SIMS, R.: Studies of arthropod-borne virus infections in Chiroptera. I. Susceptibility of insectivorous species to experimental infection with Japanese B and St. Louis encephalitis viruses. Amer. J. trop. Med. Hyg. *12:* 800–814 (1963).

SULKIN, S. E.; ALLEN, R., and SIMS, R.: Studies of arthropod-borne virus infections in Chiroptera. III. Influence of environmental temperature on experimental infection with Japanese B and St. Louis encephalitis viruses. Amer. J. trop. Med. Hyg. *15:* 406–417 (1966a).

SULKIN, S. E.; ALLEN, R.; SIMS, R. A.; KRUTZSCH, P. H., and KIM, C.: Studies on the pathogenesis of rabies in insectivorous bats. II. Influence of environmental temperature. J. exp. Med. *112:* 595–617 (1960).

SULKIN, S. E.; ALLEN, R.; SIMS, R., and SINGH, K. V.: Studies of arthropod-borne virus infections in Chiroptera. IV. The immune response of the big brown bat *(Eptesicus fuscus)* maintained at various environmental temperatures to experimental Japanese B encephalitis virus infection. Amer. J. trop. Med. Hyg. *15:* 418–427 (1966b).

SULKIN, S. E.; BURNS, K. F.; SHELTON, D. F., and WALLIS, C.: Bat salivary gland virus. Infections of man and monkey. Texas Rep. Biol. Med. *20:* 113–127 (1962).

SULKIN, S. E. and GREVE, M. J.: Human rabies caused by bat bite. Texas State J. Med. *50:* 620–621 (1954).

SULKIN, S. E.; KRUTZSCH, P. H.; ALLEN, R., and WALLIS, C.: Studies on the pathogenesis of rabies in insectivorous bats. I. Role of brown adipose tissue. J. exp. Med. *110:* 369–388 (1959).

SULKIN, S. E.; SIMS, R., and ALLEN, R.: Studies of arthropod-borne virus infections in Chiroptera. II. Experiments with Japanese B and St. Louis encephalitis viruses in the gravid bat. Evidence of transplacental transmission. Amer. J. trop. Med. Hyg. *13:* 475–481 (1964).

SULKIN, S. E.; SIMS, R., and ALLEN, R.: Isolation of St. Louis encephalitis virus from bats *(Tadarida b. mexicana)* in Texas. Science *152:* 223–225 (1966c).

SULKIN, S. E.; WALLIS, H. C., and DONALDSON, P.: Differentiation of Coxsackie viruses by altering susceptibility of mice with cortisone. J. infect. Dis. *91:* 290–296 (1952).

SULLIVAN, T. D.; GRIMES, J. E.; EADS, R. B.; MENZIES, G. C., and IRONS, J. V.: Recovery of rabies virus from colonial bats in Texas. Publ. Hlth Rep., Wash. *69:* 766–768 (1954).

TAYLOR, R. M.; WORK, T. H.; HURLBUT, H. S., and RIZK, F.: A study of the ecology of West Nile virus in Egypt. Amer. J. trop. Med. Hyg. *5:* 579–620 (1956).

TENBROECK, C.; HURST, E. W., and TRAUB, E.: Epidemiology of equine encephalomyelitis in the eastern U.S. J. exp. Med. *62:* 677–685 (1935).

TESH, R. B. and ARATA, A. A.: Bats as laboratory animals. Hlth Lab. Sci. *4:* 106–112 (1967).

TESH, R. B.; PERALTA, P. H., and JOHNSON, K. M.: Ecologic studies of vesicular stomatitis virus. I. Prevalence of infection among animals and humans living in an area of endemic VSV activity. Amer. J. Epidem. *90:* 255–261 (1969).

THOMPSON, R. D.; MITCHELL, G. C., and BURNS, R. J.: Vampire bat control by systemic treatment of livestock with an anticoagulant. Science *177:* 806–808 (1972).

TORRES, S. y QUEIROZ LIMA, E.: A raiva e sua transmissao por morcegos hematophagos infectado naturalmente. Rev. Dep. nac. Prod. Anim. *2:* 1–55 (1935).

TORRES, S. y QUEIROZ LIMA, E.: A raiva e os morcegos hematophagos. Rev. Dep. nac. Prod. Anim. *3:* 165–174 (1936).

TUNCMAN, Z. M.: Les recherches du virus rabique chez les chauves-souris en Turquie. Microbiol. dergisi Turk *11:* 80 (1958).

UHR, J. W. and FINKELSTEIN, M. S.: Antibody formation. IV. Formation of rapidly and slowly sedimenting antibodies and immunological memory to bacteriophage ⌀X174. J. exp. Med. *117:* 457–477 (1963).

VAUGHAN, T. A.: The skeletal system; in WIMSATT Biology of bats, vol. I, pp. 98–138 (Academic Press, New York 1970).

VEERARAGHAVAN, N.: A case of hydrophobia following bat bite. Scientific report, p. 40 (Pasteur Inst., Coonor, South India 1955).

VENTERS, H. D.; HOFFERT, W. R.; SCATTERDAY, J. E., and HARDY, A. V.: Rabies in bats in Florida. Amer. J. publ. Hlth *44:* 182–185 (1954).

VERANI, P.: Personal commun. (1968).

Vital Statistics of the United States: Morbidity and mortality weekly report, vol. 8, Bull. No. 38 (Government Printing Office, Washington 1959).

WANG, S. P.; GRAYSTON, J. T., and CHU, l. H.: Encephalitis on Taiwan. V. Animal and bird serology. Amer. J. trop. Med. Hyg. *11:* 155–158 (1962).

WERSCHING, S. und SCHNEIDER, L. G.: Ein weiterer Fall von Tollwut bei einer Fledermaus in Hamburg. Berl. Münch. tierärztl. Wschr. *82:* 293–295 (1969).

WERTHEIMER, E.: Glycogen in adipose tissue. J. Physiol., Lond. *103:* 359–366 (1945).

WHITNEY, E.: Serologic evidence of group A and B arthropod-borne virus activity in New York State. Amer. J. trop. Med. Hyg. *12:* 417–424 (1963).

WILLIAMS, M. C.; SIMPSON, D. I. H., and SHEPHERD, R. C.: Studies on viruses in East African bats (Chiroptera). II. Virus isolation. Zoonoses Res. *3:* 141–152 (1964).

WIMSATT, W. A. (ed.): Biology of bats, vol. I (Academic Press, New York 1970a).

WIMSATT, W. A. (ed.): Biology of bats, vol. II (Academic Press, New York 1970b).

WINKLER, W. G.: Rabies in the United States, 1951–1970. J. infect. Dis. *125:* 674–675 (1972).

WITTE, E. J.: Bat rabies in Pennsylvania. Amer. J. publ. Hlth *44:* 186–187 (1954).